Social History of Africa

PRIDE OF MEN

Recent Titles in
Social History of Africa Series
Series Editors: Allen Isaacman and Jean Allman

Cotton, Colonialism, and Social History in Sub-Saharan Africa
Allen Isaacman and Richard Roberts, editors

In Pursuit of History: Fieldwork in Africa
Carolyn Keyes Adenaike and Jan Vansina, editors

Cotton Is the Mother of Poverty: Peasants, Work, and Rural Struggle in Colonial Mozambique, 1938–1961
Allen Isaacman

Drink, Power, and Cultural Change: A Social History of Alcohol in Ghana, c.1800 to Recent Times
Emmanuel Akyeampong

Gender, Ethnicity, and Social Change on the Upper Slave Coast: A History of the Anlo-Ewe
Sandra E. Greene

Marriage in Maradi: Gender and Culture in a Hausa Society in Niger, 1900–1989
Barbara M. Cooper

TANU Women: Gender and Culture in the Making of Tanganyikan Nationalism, 1955–1965
Susan Geiger

Liberating the Family? Gender and British Slave Emancipation in the Rural Western Cape, South Africa, 1823–1853
Pamela Scully

People Are Not the Same: Leprosy and Identity in Twentieth-Century Mali
Eric Silla

A Lion Amongst the Cattle: Reconstruction and Resistance in the Northern Transvaal
Peter Delius

White Farms, Black Labour: Agrarian Transition in Southern Africa, 1910–1950
Alan Jeeves and Jonathan Crush, editors

A Green Place, A Good Place: Agrarian Change, Gender, and Social Identity in the Great Lakes Region to the 15th Century
David Lee Schoenbrun

Making Ethnic Ways: Communities and Their Transformations in Taita, Kenya, 1800–1950
Bill Bravman

PRIDE OF MEN

IRONWORKING IN 19TH CENTURY WEST CENTRAL AFRICA

Colleen E. Kriger

HEINEMANN
Portsmouth, NH

JAMES CURREY
Oxford

DAVID PHILIP
Cape Town

Heinemann
A division of Reed Elsevier Inc.
361 Hanover Street
Portsmouth, NH 03801-3912

James Currey Ltd.
73 Botley Road
Oxford OX2 0BS
United Kingdom

David Philip Publishers (Pty) Ltd.
PO Box 23408
Claremont 7735
Cape Town, South Africa

Offices and agents throughout the world

ISBN 0–325–00107–3 (Heinemann cloth)
ISBN 0–325–00106–5 (Heinemann paper)
ISBN 0–85255–682–9 (James Currey cloth)
ISBN 0–85255–632–2 (James Currey paper)

British Library Cataloguing in Publication Data

Kriger, Colleen E.
Pride of men : ironworking in 19th-century West Central Africa.—(Social history of Africa)
1. Ironwork—Africa, Central—History—19th century
2. Iron and steel workers—Africa, Central—History—19th century
I. Title
669.1'0967'09034

ISBN 0–85255–632–2 (James Currey paper)
ISBN 0–85255–682–9 (James Currey cloth)

Library of Congress Cataloging-in-Publication Data

Kriger, Colleen E.
Pride of men : ironworking in 19th century West Central Africa / Colleen E. Kriger.
p. cm.—(Social history of Africa, ISSN 1099–8098)
Includes bibliographical references and index.
ISBN 0–325–00107–3 (alk. paper).—ISBN 0–325–00106–5 (pbk. : alk. paper)
1. Ironwork—Africa, Central—History. 2. Blacksmithing—Africa, Central—History. 3. Ironworkers—Africa, Central—history.
I. Title II. Series.
TS304.A352K75 1999
669'.1'0967—dc21 98–30625

Paperback cover photo: Blacksmith Ikete Boteni making an arrowhead, Lopanzo, Equateur, Zaire, 25 July 1989. Photograph by the author.

Printed in the United States of America on acid-free paper.
02 01 00 99 BB 1 2 3 4 5 6 7 8 9

To my father

Contents

ILLUSTRATIONS

Acknowledgments

Some portions of this book appeared earlier in the form of a Ph.D. dissertation in history, and so I first acknowledge the many encouragements and guidance given me by my committee members: the late Dr. Zdenka Volavka, Drs. Paul Lovejoy, A.S. Kanya-Forstner, Martin Klein, and Ted Bieler. They have supported my project and me over the years with patience and good humor, even though measuring knives and hatchets might have seemed on occasion to be a dubious enterprise. When Dr. Volavka died, I thought I would never finish; Paul Lovejoy stepped in as supervisor and led me over the final hurdles of the Graduate History Program and out the doors of York University. I also owe a very great debt to several other experts and master ironworkers whose tutelage and help have been indispensable: J.E. Rehder, Lotoy Lobanga, David Norrie, and Daniel Kerem.

Working in Lopanzo would not have been possible without the generous cooperation of Drs. Eugenia Herbert and Kanimba Misago, who allowed me to join their research team in 1989. All of us remain ever grateful to Lema Gwete, Père Vinck, Boilo Mbula, and Kathryn Delaney for providing the logistical support on which our projects depended. I personally thank Nkumo Bipa, Amba Mbili, and Balaka Bolangondo for assisting me as interpreters.

Curators and staff of museums in North America, Africa, and Europe graciously opened up their collections for me, and accommodated my research needs in spite of the inconveniences they often posed. Dr. Gertrude Nicks and Mary Hayes of the Royal Ontario Museum were especially patient and helpful during my weekly visits to their department over a period of several years. I owe many thanks to individuals in three departments of the Musée Royal de l'Afrique Centrale: Mme. Huguette Van Geluwe, Dr. Marie-Louise Bastin, and Dr. Gustaaf Verswijver in Ethnology; Dr. Cl. Gregoire of Linguistics; and Dr. Ph. Marechal and Patricia Van Schuylenbergh in History.

Drs. John Mack and Christopher Spring of the Museum of Mankind in London kindly made available to me not only the metalwork in the Torday collection, but also the photographs, field notes, and correspondence from the Congo expedition. I was fortunate to have access to collections in the National Museum of Natural History in Washington, D.C., thanks to the cooperation of Dr. Mary Jo Arnoldi, Susan Crawford, and Felicia Pickering. Mary Lou Hultgren made it possible for me to work closely with the Sheppard Collection at Hampton University, and she and her staff made it a pleasant experience as well. Dr. Phillip Lewis and Christine Gross of the Field Museum of Natural History in Chicago helped me survey and work with their collections; Dr. Enid Schildkrout of the American Museum of Natural History in New York assisted me with the Lang collection of photographs; and Père Vinck generously shared with me the collections at Centre Aequatoria in Zaire.

Staff and archivists at the Regions Beyond Missionary Union, London, and the Archives Africaines, Brussels, were particularly courteous and helpful, especially Dr. Françoise-Marie Peemans and Mme. Dukait. I cannot adequately thank Mary Lahane and Gladys Fung of the Interlibrary Loan Office at Scott Library, York University, and Gaylor Callahan of the Interlibrary Loan Office at Jackson Library, University of North Carolina at Greensboro, for never giving up on and never being annoyed at my requests.

I always envisioned this project as a long term one, and so many agencies and institutions have supported it over the years. Looking back on all the grant applications and competitions still fills me with wonder that I was able to keep afloat and bring the project to fruition. I am thankful to have received funding from York University, Ontario Graduate Scholarships, the Queen Elizabeth II Ontario Scholarship, the Social Science Research Council, Fondation Dapper, the Social Sciences and Humanities Research Council of Canada Post-Doctoral Award, the Kohler Fund and Research Awards from the University of North Carolina at Greensboro, the Zdenka Volavka Postdoctoral Research Fellowship, and the American Philosophical Society. I must also acknowledge the support I had during graduate school from the decent assistantship salaries that had been negotiated over the years by the Canadian Union of Educational Workers and York University, and the relatively progressive taxation policies of the Canadian Government, which allowed me to deduct research expenses before paying tax on research income.

Two people deserve special mention for having been steadfast, inspiring, and invaluable mentors over the years, each in their own inimitable way. Zdenka Volavka gave me more than I can measure—as a historian and as a human being—and I will always miss her, above all her fierce skepticism, her deep respect for the past, and her wicked sense of humor. Bringing this project to a close might not have been possible had I not been encouraged all along the way by Jan Vansina. Crowning our intellectual exchanges was the postdoctoral year I spent in Madison, when he initiated me into the worlds of historical linguistics and comparative Bantu, and shared with me countless details of central African history. Both of these people embody what the scholarly enterprise is ideally intended to be.

I was fortunate to have feedback, advice, and criticism from several readers. I thank Jean Allman, Robert Harms, Allen Isaacman, David Killick, and Jan Vansina, and an anonymous reader, though I take full responsibility for the text as it now stands. Stuart Marks, Kalala Ngalamulume, and Adria LaViolette may be surprised to know this, but they each buoyed my spirits at crucial moments. I thank Dr. Jeff Patton, Geography Department, University of North Carolina-Greensboro, for helping me with my base map, and Chris Schierkolk for his courtesy and cooperation in creating each of the maps included in the book. I am grateful to Tim Barkley and Dan Smith, Creative Services, UNCG, for their help with photographs.

My very special, closest friends have brought joy to my life over and over again in countless ways. I am honored to know them. Thanks to Betsy Eldredge, Oded Frenkel, Françoise Grossen, Frank Melton, Adrienne Middlebrooks, Elaine Naylor, Ann O'Hear, Wendy Thomas, and Lisa Tolbert. My dear father taught me all about troubleshooting—taking small engines apart and putting them back together—and so it is to his memory that I dedicate this book.

A Note on Orthography and Toponyms

Words from Bantu languages are reproduced here in their forms as published or transcribed, that is, with tones when they were originally given, without tones when they were not. The distinction between the different "e" and "o" vowels (open and closed) is not reflected in the transcription. I use the terms "Zaire" and "Zaire river" throughout the text, as these have been current for the past two decades and are well entrenched in the scholarly literature. To use the term "Congo" instead would invite confusion between the various entities that have been known by that same name, e.g., French Congo or Belgian Congo. My choice is one based strictly on clarity and convenience, and carries absolutely no political implications. It should go without saying that I do not in any way endorse the former Mobutu regime or the policies of "authenticity" that he instituted.

Glossary of Technical Terms

Annealing Alternately heating and cooling metal slowly to make it tougher and less brittle (see Tempering).

Anvil A solid mass of stone or metal, shaped with a working surface suitable for hammering out metal into objects.

Bellows An instrument designed to pump or force air into a fire or furnace.

Bloom The heterogeneous mass or chunks of iron metal produced by an iron smelt, particularly at the stage before refining out most unwanted impurities such as slag and charcoal.

Bowl furnace A type of smelting furnace made by digging out a curved hole in the ground.

Burin An instrument with a sharp point or cutting edge, designed for incising and engraving.

Cast iron Iron that has reached liquid form by melting, which is then poured into a mold; iron with a carbon content of about 2 to 4.5 percent.

Chasing Technique of embellishing a metal surface by cutting lines or grooves into it and/or by creating relief patterns with raised work.

Chisel An instrument with a sharp cutting edge, often beveled, designed for cutting and carving a softer material.

Engraving Technique of embellishing a surface by cutting and/or carving out lines and shapes.

File An instrument with a very hard, roughly textured surface area which can be used to smooth out and polish an object.

Filigree A technique of embellishment or decoration, often very finely and delicately executed, using units of metal wire twisted, wound, or plaited together.

Flux Material that enables the flow of other materials to occur or to occur more smoothly, such as material added to a furnace to enhance the process of smelting ores, or material placed between two surfaces in welding to improve the quality of the weld or joint.

Forced draft Air channeled into a furnace, especially by using bellows.

Forge A fireplace where a blacksmith works metals by heating and hammering; a smithy or blacksmith's workshop.

Foundry A workplace where furnaces operate to smelt ore, and/or where metals are melted and poured into molds to make ingots or cast artifacts.

Gangue Ordinary materials that are found intermixed with useful minerals in ores; materials that become waste runoff during the smelting process.

Haft Pointed termination located at the base of a blade, allowing it to be inserted into a handle or shaft.

Hammer An implement held in the hand for delivering percussive blows to another object.

Hematite (Fe_2O_3); a mineral ore of reddish color; the major source for smelting iron ore into iron metal.

Laterite A reddish soil found in the tropics, often with sufficient mineral content to be used as an ore for smelting (e.g. iron, manganese, aluminum, etc.).

Limonite [$FeO(OH)\ nH_2O$]; a hydrated iron oxide, sometimes called bog iron, that is very commonly found in damp soils; a form of laterite.

Microstructure What a microscope shows to be the underlying but (to the naked eye) invisible construction or organization of a material.

Midrib A rib or spine that extends along the central longitudinal axis of a metal blade, mainly to strengthen it.

Natural draft Air supply into a furnace, occurring without bellows.

Ogee midrib A form of midrib that creates a double curve or zigzag along the horizontal cross-section of a blade.

Pig iron Raw cast iron from a blast furnace, molded into ingot forms.

Punch An implement designed to create holes in an object, or various stamped patterns on its surface.

Quenching Technique used to harden high carbon iron by heating it, then plunging it into a liquid to quickly cool it.

Reduce To remove the oxygen and impurities from an oxide or ore, accomplished by smelting.

Shaft furnace A type of furnace design that includes some form of vertical structure or walls, sometimes placed over a pit or hole in the ground.

Slag Waste materials that have melted and flowed out of mineral ore during a smelt.

Slag-tapping Design feature of a smelting furnace for allowing melted slag to run off and away from the working area.

Smelting To heat and reduce ores, so that the metal is separated out from other materials in the ore.

Smithing To form and shape metal by heating it and hammering it.

Socket A tubular termination at the base of a metal blade, so that a shaft may be inserted into it.

Stamping Technique for decorating or marking a surface with graphic imagery or hammered relief patterns.

Steel A term often used imprecisely to refer to iron that is tougher and harder than wrought iron; more precisely, alloys of iron and carbon that show a carbon content higher than wrought iron and lower than cast iron, usually between .25 and 1.75 percent.

Tempering Technique for strengthening a metal by alternately heating and cooling (see Annealing).

Tuyère A tubular pipe or opening which allows air to enter or be forced into a furnace or fireplace to control the temperature.

Welding To fix two metal surfaces together either with heat and pressure only, or with heat and a material used as a flux or solder.

Wrought iron A term used generally to refer to alloys of iron and carbon that are soft and malleable, with carbon content less than about 0.1 or 0.15 percent and with slag and other impurities of about .5 to 1.0 percent.

ABBREVIATIONS

Acc.#	Museum accession number
AMNH	American Museum of Natural History, New York City, USA
FM	Field Museum of Natural History, Chicago, Illinois, USA
HUM	Hampton University Museum, Hampton, Virginia, USA
IMNZ	Institut des Musées Nationaux du Zaire, Kinshasa, (formerly) Zaire
JAH	Journal of African History
MOM	Museum of Mankind, London, England
MRAC	Musée Royal de l'Afrique Centrale, Tervuren, Belgium
NMNH	National Museum of Natural History, Washington, D.C., USA
ROM	Royal Ontario Museum, Toronto, Canada

Bantu language groups are referred to according to Guthrie's classifications, as revised by the research team at the Section Linguistique, Musée Royal de l'Afrique Centrale, Tervuren, Belgium [e.g. (K23), (C61), etc.].

PART I

THE WORK AND THE SETTING

1

African History, Ironworking, and the Mystique of the Blacksmith

"For pride—pride of men" is a phrase that was repeated to me many times in the Zairian town of Lopanzo, when I asked why certain types of extravagantly crafted luxury products had been made by blacksmiths of the past. A similarly opaque notion underlies the generic references to many such products—elaborate ceremonial knives, hatchets, and other costly metal implements—which were known as "swords of honor," "swagger hatchets," or "emblems of dignity" throughout the region of the Zaire and Kasai River basins in west central Africa. During the reconstruction of an iron smelt in Lopanzo I witnessed more obvious and direct evidence of pride in ironworking itself. This was displayed sporadically when elders took care to instruct the younger generation of men in the town how to recognize which iron ore they thought was of higher quality, for example, or when townspeople roared in unison to celebrate the success of the smelt. Pride was again in evidence when elders brought out smithing tools and other masterworks of former smiths from treasuries for ritual and for show. In contrast with other important historical relics, these were never offered for sale. This book represents my search to explain the social and economic values of ironworking in nineteenth-century west central Africa—how, why, and in what ways it was an occupation created by and for "pride of men."

This investigation centers on the ways in which ironworkers throughout the Zaire River basin built a craft tradition and a distinct and admirable identity for themselves in their societies. That they did so is important to our understanding of the roles of artisans in precolonial central Africa and their positions as founders and shapers of workshops, labor forces, and communities in rural and urban settings; as innovators in social and economic networks; and as a class of individuals deemed worthy of respect. How they built this tradition and identity is also important to our understanding of precolonial central African history, and why blacksmiths were so admired and have figured so prominently in the orature of the region.

My argument consists of two main strands. First, I stress that this identity can only be appreciated when the work itself is described and analyzed systematically. I begin by demonstrating that the economic expansion of the nineteenth century (ca. 1820–1920) saw the florescence of a craft tradition that had been developing over two millennia. Most of the book is then devoted to examining ironworking in detail during this crucial hundred-year period. My evidence shows great variety and flexibility on the part of ironworkers at that time, not only in what they produced but also in where, how, and for whom they produced it. Such maneuverability suggests that over the course of its development, ironworking had been maintained and modified in accordance with contingent, historical factors, and above all, by the interests and actions of the ironworkers themselves.

Systematic analysis has allowed me to identify several important features of ironworking that clearly affected the status of the workers. One of these was the division of the occupation into two separate but interrelated specializations, smelting and smithing, each with its own technologies and work patterns. I show that the production levels of nineteenth-century smelters were, for a number of reasons, generally low in relation to potential demand. As a result, iron was valued very highly and sustained demand for it generated an extensive trade in all sorts of iron products, including semifinished goods that circulated as currency. The second important feature was the high labor status of those who entered the occupation. What my evidence shows convincingly is that ironworkers in both smelting and smithing specializations maintained a considerable degree of control over their workplaces and how their work was organized, and hence could exercise a great deal of autonomy. This shared autonomy did not make all of them equal, however, which brings us to the third feature of the work. Masters created and maintained hierarchies within the occupation according to skill, and it was masters, the most highly skilled, who were most likely to enjoy privileged status. Moreover, it was their drive to maintain these hierarchies that encouraged product innovation and technology transfer from one workshop to another. The fourth and final feature has to do with master blacksmiths who, at least during the nineteenth century, were the most successful and well-rewarded ironworkers, and the ones who were remembered by later generations. Not surprisingly, my analysis of their work shows that the contribution blacksmiths made to hunting and agricultural production was very important. But this alone does not explain why they were considered so valuable to society. Smiths reached into many other societal domains as well; they made all sorts of metal currency units, along with the impressive luxury products mentioned earlier, and a good deal else.

I emphasize throughout that the prominence enjoyed by ironworkers in the nineteenth century, while stemming in part from the smelting work they did at the furnace, was primarily the result of their smithing, the complex day to day work that they did at and around the forge. This particular side of ironworking, the making of metal goods—especially as it was practiced in central Africa prior to the twentieth century—has been under researched and hence poorly understood by scholars. With regard to the work of blacksmiths, two major lacunae have existed in the literature. First, it was not known how numerous the forms of iron currencies were in the nineteenth century, nor had the complexity of their zones of circulation been acknowledged or appreciated. Second, the regional luxury markets for metalwares had not been identified or delineated fully, let alone analyzed to reveal changing production and consumption patterns. By making a start at filling in these lacunae, I am able to describe more completely just what blacksmiths' work was, and how it changed as they invested in their own retraining and upgrading of skills and market share. This in turn contributes to our understanding of how and why they were so respected and admired in the region.

The second strand of my argument takes up various social issues. As I survey a range of ironworkers' workshops and strategies, we begin to see how master smelters and smiths wielded, adhered to, and transgressed social tools, norms, and categories as they shaped and reshaped their craft tradition and group identity. As autonomous workers, masters were able to structure and manage the labor processes, the complex webs of tasks and social relations involved in production. In doing so, they used their positions as heads of households and families to recruit dependents as unskilled or semiskilled workers in order to intensify the work. Masters also incorporated religious ceremonies and ritual to consecrate the workplace at times when particularly difficult, risky, and momentous tasks were being carried out. In these and other ways, ironworkers defined their work according to locally based ideological norms and conditions, and so, on these occasions, upheld the *status quo*.

I argue that ironworkers were, on other occasions, agents of change. In pursuing their careers, they persistently transcended local and familial affiliations, crossed ethnic and language boundaries, and generated a regional dimension to their work and group identity. Masters promoted themselves and used their reputations to generate social ties with other ironworkers or with potential patrons far away, transferring skills, techniques, and knowledge to new locales. As they did so, they could enter the circle of regional elites formed on the basis of wealth and privilege. Workers were not formally organized beyond their workplaces, but neither were they disorganized or confined to their communities by kinship and ethnic ties. What united them regionally was their social prominence and gender. Absolutely no evidence has come forth that there were any women who were smelters or smiths, and there is a great deal of evidence—at least for recent times—that women of childbearing age were prohibited from coming near the workplace or consorting with workers at certain times.[1] In short, what made ironworkers' status special came from their diligence as social beings who used their autonomy in a variety of ways to organize and legitimize their work, to develop regional networks and relationships, and to define their occupation and their privilege as male prerogatives.

The aim here is to present a systematic and thorough study of ironworking in west central Africa, a region where it is well known (though only generally) that blacksmiths were important historical figures.[2] My evidence offers no single model, no preeminent single factor that would account for the prominence of ironworkers, individually or as a group. In arguing why this should be the case, I emphasize ironworkers' autonomy and agency as they traversed social categories, and their ubiquity as the products and impacts of their work reached into many social domains. My focus is primarily on the workers and the work, showing them in full array, with their flexibility and diversity intact. Ironworkers, while making their occupation and status, both fitted into and used the social organization and structures of their times, but were not wholly molded by them.

THE WORK

Ironworkers in nineteenth-century central Africa were maintaining a craft tradition that went back over two millennia. However, the earliest evidence for this is sparse and uneven, hence in some ways very misleading. Research in historical linguistics has not yet delineated any coherent patterns for the development of expertise in ironworking among early Bantu-speaking peoples. Much of the archaeological research so far has been focused on searching for evidence of the beginnings of iron smelting and its relationship with copperworking. Given this paucity of data, it is easy to overlook a major gap in what evidence there is: relative to smelting, smithing has been almost entirely neglected. This can lead to the false impression that ironworkers concerned themselves mainly with primary manufacturing, the making of iron metal.

It is important then, to begin here with a brief description of just what ironworking entails. For although it is often considered to be a single occupation, it includes two discrete specializations: the independent yet interrelated working processes of smelting and smithing. These are central to understanding and tracing ironworkers' histories, since the basic material and technical requirements of each one served as the armatures around which central African smelters and smiths built their complex technologies and workshops. We will first focus on just what it is that constitutes these armatures, that is, the ergonomics or essential components of the two working processes. It will then be possible to see how different ironworkers developed their own variations from place to place and made changes to their work over time.

In order to work with iron it must first be extracted from its ores. Since it binds readily to oxygen, iron is hardly ever found already in metallic form. Instead, it must be "won" from iron ores, many of them rather unobtrusive, humble-looking rocks, and there is no single, simple way of doing this. It is the process of smelting that turns ore into metal. During smelting, chemical reactions take place between the ore and the fuel in a furnace at temperatures above the melting point of the metal involved. Modern industrial smelters produce cast (molten) iron by smelting it at very high temperatures in large blast furnaces. But this is only one way of producing iron. Smelters can also produce iron in the solid state by reducing the iron oxides to iron metal at temperatures below the melting point of iron (1538°C). It is the former method that is presently most familiar to Europeans and North Americans, while it is the latter that was employed by ironworkers in Africa.

In the solid state process of smelting, the fuel used in the smelting furnace is almost always charcoal. The charcoal not only provides heat but also combines with oxygen in the air blast to make carbon monoxide, which in turn reduces iron oxides, producing iron metal and carbon dioxide. It is the carbon monoxide that is critical, for it provides a reducing atmosphere for smelting iron ore and, by limiting the amount of air supply to the furnace, smelters can create more carbon monoxide and thus a larger working area in the furnace. As the oxygen is reduced from the ore, other molecules of carbon may combine with the iron, and the lesser or greater amounts of this combined carbon result in iron with different properties. Low carbon iron is soft and malleable; iron with moderate carbon content (steel) is durable and tough; very high carbon cast iron can be brittle and very difficult to forge. Also during the smelt, minerals such as quartz and clay, which are collectively called gangue, separate out and run off in liquid form as slag. Thus smelting in the solid state involves a delicate balancing of the important factors of air supply and temperature. The air supply must be kept low enough to keep the heat below the melting point of iron and to create enough carbon monoxide to reduce the oxygen from the ore. It must also be high enough to reach the temperature needed to make a fluid slag (about 1100°C) that will drain away from the metal. If all goes well, the furnace will yield clumps or a mass of reduced iron at the end of the smelt. This iron is called a "bloom," and it usually requires refining, that is, the hammering out of trapped lumps of slag and charcoal. Finally, the end product is iron metal that can be hammered and formed into products by the blacksmith.[3]

Producing blooms, as challenging and important as it is, is therefore just the beginning, an early step in a series of ironworking processes. Iron must now be worked into marketable products. This is done by smithing, the manipulation by hammering of workable iron. The main tools used for this are the anvil and the hammer. The anvil is a hard surface of stone or iron on which the iron metal is placed and turned. The hammer, which can be made of either stone or iron, is used to "draw out" the metal by stages into the desired form. The smithy or forge also includes a charcoal fire kept hot with air pumped through bellows. In the fire the smith heats iron to the desired temperature, then hammers it, reheats it, hammers it again, over and over. Smithing is done to further refine a bloom of iron from the furnace; unwanted slag and charcoal bits are hammered out, and the portions of malleable iron remaining are then welded together until the bloom becomes a coherent mass of metal. Smithing can stop here, with the refined iron stored for later use or traded to other smithies or consumers. This workable iron subsequently becomes a semifinished or finished product by further drawing out, that is, by heating, hammering, and frequently turning the iron on the anvil to receive the hammer blows from different directions. The complexity of the product's form depends upon the level of skill of the blacksmith.

Smithing, the making of products out of iron metal, comes after the smelting of iron ore in the production process, but historically it usually comes first. Although the precise historical conditions of ironworking beginnings were undoubtedly variable, it is unlikely that a society would become an iron-producing one all at once or at the outset. As archaeologists well know, it is much more likely that societies would become first iron-using ones through trade, and that blacksmiths would begin to work iron initially by repairing or reworking imported goods. It is therefore safe to assume that more central African ironworkers were blacksmiths rather than smelters. In other words, although more research attention has been given over to the study of iron smelting, the work of smithing has gone on for a longer period of time and has involved a greater number of workshops and specialists. The occupation of ironworking involved both primary manufacturing and secondary manufacturing. In specific historical times and circumstances, some of the workers were smelters, most of them were smiths, and many combined both.

Furthermore, complex specialization patterns resulted from the various training opportunities and work strategies followed by the workers. If the evidence gathered in this study can be relied upon to suggest lines of technology transfer in the deeper past, smelting expertise was usually transferred to new areas where blacksmiths already existed. The desire to learn smelting technology ought to be viewed as a strategy adopted by blacksmiths for import replacement, providing themselves with more reliable and alternative sources of iron metal. Seen in this light, ironworkers that engaged in both smithing and smelting were not unspecialized but were diversifying their operations.

AFRICAN IRONWORKERS IN THE SCHOLARLY LITERATURE

While I was honing in so closely on the complexity and diversity of ironworking itself, several distortions and misunderstandings in the literature came to light. One of these is an emphasis that has been placed on occupational "castes." For example, blacksmiths in and around the Mande-speaking areas of West Africa have been the best known ironworkers in the scholarly literature. What has seemed most striking about them is their very particular social status, that is, that they were viewed with ambivalence by others in their societies. Smiths were feared and sometimes despised, and at the same time acknowledged as powerful and skilled. Their status was shaped by the so-called "caste" system of which Mande smiths were a part, and only recently have scholars begun to see this system in historical terms.[4] Disentangling the actual work of the occupation from the social institution of the endogamous "caste" becomes a problematic task, and it is not at all clear yet how or why they became entangled in the first place. What is most important for our purposes here is that no such "caste" system arose in west central Africa, and so the quite different social status of ironworkers there serves as a useful comparison.[5]

Other distortions occur when ironworkers of a particular time and place are treated as representative examples of ironworkers in general. The Mande case has loomed large in historians' minds as a standard by which other African ironworkers are often measured and assessed, even though very little is known about ironworking in precolonial times.[6] As an example, important questions surrounding the roles played by Mande blacksmiths in "power associations" are not at all resolved historically. McNaughton thoroughly explored this issue as it pertains to the present century, basing his work primarily on evidence from oral testimonies and traditions. He mentions a number of these social institutions, which exercised judicial and policing authority, their religious sanction coming from supernatural forces of various kinds. He focuses on one in particular, *kòmò*, an institution monopolized by smiths. But as McNaughton himself acknowledges, it is still very uncertain just how old this smiths' "power association" and its ideological paraphernalia actually are.[7] Hence the elegant argument constructed recently by Brooks, in which he has blacksmiths carrying their *kòmò* association southward with the spread of Mande traders and settlers during a dry climatic period from 1100 to 1500, sounds plausible but has not actually been proven.[8] What we have so far in the Mande case is mainly twentieth-century evidence, evidence that cannot be projected back in time or onto other societies on the continent. Much is indeed known about recent Mande blacksmiths, but their histories over the longer term, along with the histories of other ironworkers elsewhere in Africa, remain to be written.

Aside from the Mande case and other less well-known anthropological case studies, much of the literature on African ironworking focuses on smelting and issues related to it.[9] A major reason for this emphasis is the unavoidable skewing of archaeological evidence in favor of furnace remains, which I discuss in Chapter 2. Compounding this skewed evidential base are the kinds of questions scholars have posed, as can be seen in several major debates and research programs over the years. One of these centers on technical questions about the iron that was smelted in African furnaces, its specific properties, and how it compared to iron imported from Europe.[10] This particular debate has not been as instructive as it could be, one of its greatest weaknesses being the neglect of blacksmiths, the ones who worked with smelted iron. All too often scholars have approached the debate with the preformed judgment that high carbon steel is always and everywhere considered to be superior to iron with a lower carbon content. My evidence for nineteenth-century west central Africa calls into question this and other assumptions about iron technology, and indicates, for example, that blacksmiths there preferred to work with low carbon iron, hardening it into serviceable tools by cold hammering.[11] Hence the debate about the "quality" of iron smelted in Africa has suffered from a failure to take into account alternate approaches to defining "high quality."

Other debates have generated rather ahistorical images of African ironworkers, images this book seeks to dispel. Questions pursued by archaeologists about the earliest patterns of settlement in central and southern Africa and questions pursued by linguists about the development of Bantu language groups inadvertently create a simplistic image of ironworking in the distant central African past.[12] Experts within these disciplines appreciate the complexities of working in iron, its social and economic implications, and the difficulties of summarizing the wealth of their evidence. But to the nonspecialist who bypasses the particulars and looks only at the streamlined results of this kind of research, a "search for origins" by archaeologists and linguists seems to suggest a highly reductive and evolutionary image of ironworking: iron is introduced to and used by a society, the technology is then transferred into that society and further developed over the long term, and the products and skills of ironworkers slowly become more specialized. In other words, the ironworker is a passive product of society, apparently the reflection of basic, long-term social and economic trends.

Finally, there is the issue of religion and ritual in ironworking, and above all in smelting processes. An image of ironworkers as shamans has been created from studies in the fields of comparative religion and structural anthropology.[13] It is an image that presents the blacksmith as a ritualist and magician-priest revered in societies all over the world presumably because of his miraculous ability to transform precious materials. But again, the image is simplistic and schematic. Such a narrow focus on the rituals performed in their workplace again portrays ironworkers as passive—this time as the reflections of locally held cultural and religious beliefs.

THE MYSTIQUE OF THE BLACKSMITH

Schematic, reductive images of ironworkers that crop up in the scholarly literature parallel in certain respects the ones presented in central African orature. Episodes marking critical junctures in the histories of several prominent central African kingdoms feature various types of blacksmith cliché. The blacksmith appears as a founding or civilizing hero and, if he himself is not the wielder of political power, then he is in one way or another the ceremonial legitimizer of it.[14] At least one historian interpreted such clichés rather literally and posited that control over iron ore and salt deposits was a determining factor in early state formation.[15] Others have seen these clichés figurally, with metalworking processes used in oral tradition as a symbolic language for embellishing pivotal leaders or periods in the past. But neither of these interpretations is wholly satisfactory, and all of the foregoing images are flawed and incomplete. Combined, they form and perpetuate what can only be called a "mystique" of the blacksmith.

What is perhaps most misleading about this "mystique" is how it seems to place the ironworker in time but outside history, that is, he appears insulated from the contingent happenings and conditions of his particular surroundings. Inhabiting the realms of deepest archaeological and linguistic time, the ironworker exists like some type of prehistoric creature caught and fixed in amber, becoming over the slow grind of centuries the relic of a very distant past. If he changes at all, it is only in tandem with general trends. Similarly, in mythological time the ironworker serves as a basic medium through which culture, ritual, and tradition pass. He appears as a figure of prehistory, cosmology, and the occult, beyond our reach, both mysterious and awe-inspiring. He belongs to the world of religious belief, ruled primarily by magical forces and mysticism, and far from the profane world of everyday affairs.

The mystique of the blacksmith tempts and fascinates, for it portrays him in idealized and romanticized forms. It should not, however, be allowed to detract from the more mundane details of his work, especially at the forge. For work is what all blacksmiths did, making metal products day after day, and bringing a historical perspective to their work can shed light on their more modest proportions and human volition. This study adds more particularity and variety to their features in order to demystify ironworkers. Above all, I seek an understanding of how they responded to and initiated historical change. Over the past millennium, as central African societies and economies grew increasingly complex and as political leaders competed for power over rising and declining chiefdoms and kingdoms, smelters and smiths were not living in a separate temporal domain. They were very much a part of that history, living right in the midst of those eventful times.

Central African ironworkers were historical figures, shaped by and shaping the particular places and time periods in which they lived. They are not interchangeable with ironworkers in West Africa, nor were those living in 900 or 1400 the same as the ones living in 1870. They were probably always important, but for different reasons and in different ways as they adapted their work to new surroundings and opportunities. In this book I seek to situate them in historical time and to show how much and in what ways they acted to maintain and change their worlds. Their history involves more than the initial introduction of iron into society, and ironworking was not a complete, self-contained package of tools and processes that was introduced once and then simply continued for centuries. That is why, for example, there is not a neat and general correlation between the distribution of Bantu language groups and the historical patterns of when, where, and by whom different ironworking processes were carried out.[16] If ironworking skills had been developed and held securely within family groups and local communities, patterns of change would have been slow and in accordance with the long-term social trends that produce new languages. Moreover, the history of ironworkers involves much more than ritual. Here I aim to present in greater detail and as fully as possible the range of actual work that the smelter and blacksmith did, in addition to the rituals that on occasion surrounded them.

IRONWORKING AS SOCIAL HISTORY

I began this research project assuming naively that in the end I would be able to turn my findings into a coherent, chronological narrative of the history of ironworking in west central Africa. Instead, what I found during the course of my research was an exhilarating but at the same time almost stupefying web of overlapping and interlocking patterns of change in the workers, their tools, products, and workshops. I uncovered many interrelated histories of ironworking, a richness that I never could have imagined beforehand, and which cannot be reduced to a single narrative line. That very complexity is itself one of my more significant findings and a common theme throughout this book. It resulted from what I see as the essence of ironworking in this particular time and place, that is, that skilled workers were autonomous and independent and so could follow various options and opportunities over their lifetimes. Along with the mixings and movings of their households and communities and the varying ways ironworkers responded to and created changes over time in their markets, the diversity of the work took shape.

What these independent workers held in common was their special social identity and place of privilege in their communities. In this special identity, which is the major finding of this book, central African ironworkers exhibited features very similar to artisans in other historical areas of the world. Their training and valuable skills set them apart from other kinds of laborers, and their work was considered by themselves and others to be indispensable. The economic success they were able to achieve garnered for them a high social status and the rewards of respect and admiration. Moreover, and what makes this particular historical example so intriguing, the artisans themselves had a hand in creating the very cultural values that nurtured their markets. It was the preciousness of iron which could make smiths, especially, so successful as the makers and converters of money and fashioners of metal luxury goods. Combined, these features demonstrate that ironworkers in precolonial central Africa, perhaps even more so than artisans elsewhere, were a crucially important stratum in their societies.

Devoting an entire study to this important group of artisans during their heyday broadens and deepens the scope of African social history. Since the 1970s, there has been steady growth in historical research on social groups in Africa. As a result, the African historiography now includes not only studies of important leaders, political conflicts, diplomacy, and forms of political organization but also how African societies have been built and shaped by, among others, slaves, food producers, merchants, porters, wage laborers, and women.[17] Research into the histories of work and labor has followed three main paths: slavery and other forms of unfree labor; export production in agriculture, especially that of cash crops; and wage labor (often migrant) in capitalist and/or industrialized economic sectors.[18] Much of it, in the latter two categories especially, deals with the twentieth century. Largely missing from this literature on work and labor in African social history is the artisan[19] —how he or she created specific forms of work, facets of social identity, and cultural values in precolonial times, and then how these were challenged by colonial and postcolonial regimes and the effects of the international economy.

In this book I analyze ironworking as an artisanal occupation and as a craft tradition—how men worked in it and how the occupation worked for men. I establish that ironworkers were a distinct social group, and also that the occupation provided individual men with social and geographical mobility. In addition to the hierarchical divisions within their numbers, ironworkers were linked together across space and time by a shared identity based on their occupational skills, gender, and privilege. Their complex social relations, from the formation of different kinds of labor processes in production to the varied commercial and consumer markets they served, have, at various times, placed ironworkers in positions almost everywhere across the social spectrum. Knowing their history thus adds new dimension to our understanding of the structures and dynamics of precolonial central African society.

Placing the work and the workers front and center as the focus of analysis allows me to show how smelters and smiths were able to operate within, between, and outside social structures. I make references throughout this study to gender, ethnicity, class, labor processes, family, household, slavery, and other historical themes and categories of social analysis, but I deliberately avoid emphasizing any one of them. For while each of these categories offers a very promising avenue for further research, to follow any single one of them now would shift attention away from the work and onto social structures, which would contradict my finding that ironworkers were autonomous. Their work was highly gendered, for example, but gender alone does not get at the core of what went on with ironworkers, or the multiple sources of their autonomy and social power. They were laterally mobile, not being bound necessarily by language barriers and ethnic affiliations,[20] and they were mobile vertically as well, inside and beyond their occupational group. Masters strove to differentiate themselves from other ironworkers by acquiring the higher levels of knowledge, skill, and managerial experience which made it possible for them to become members of wealthy and powerful elites. Ironworkers were flexible, too, varying the labor processes in manufacturing to suit specific and changing conditions. The family, household, and slavery all could come into play at one time or another, as pools of labor and dependent relations to be exploited, but none of these categories defines, once and for all, the production organization of smelting and smithing.

Indeed, adopting the narrow perspective of a single social category would deny an important feature of central African ironworking in the nineteenth century, which is the fact that it permeated so many aspects of peoples' lives and values. It was part of the material base of society. Those who had direct and sustained access to this well-established form of wealth grew richer, increasing economic inequality and social stratification across the region. Gender inequities were exacerbated as well, as gendered differences in access to ironworkers' technical knowledge and special skills created barriers to womens' direct access to iron. In other words, ironworkers could revise and restructure social organization. Searches by them for better, more stable workshop locations and proximity to iron ore could lead to the creation of entirely new communities, as examples from the Ngiri basin, the middle Zaire, and other areas show. Ironworkers also reshaped already existing communities in various ways. When they left their own homes to resettle elsewhere, and when they became immigrants in other villages and towns, they changed the social, economic, and cultural complexity of those populations. The example in Chapter 5 of Sala Mpasu smiths becoming members of a Kete polity demonstrates this point very well. Ironworkers reshaped population densities, too, by investing their wealth in women and slaves and by attracting apprentices, other ironworkers, and patrons. And throughout the region, blacksmiths made many visual markers of the changing social and official relationships that were the structural elements of central African communities.

Smelters produced wealth from ore and blacksmiths distributed it by choosing to form it into currency, tools, fancy metalwares, or other types of products. That is why histories of the objects they made are such an important component of the analysis presented in this book. These objects offer entry into the ironworkers' social, cultural, and working worlds, demonstrating along the way how material sources and lexical data can be particularly appropriate and useful primary sources for research in social history.

SOURCES, METHODOLOGY, AND CHRONOLOGY

To illustrate this point, I will briefly cite two ironworkers' histories that are also discussed later on in the text; one is derived from the Kuba *ikul* knife, and the other, from luxury hatchets made at Luluabourg workshops. In piecing together these histories of ironworkers, I started with technical and comparative analyses of iron products—including currencies, tools, and prestige goods—drawing from them logical inferences about their makers. After that, I sought corroboration from other sources such as lexical data, contemporary accounts, and oral testimonies. The first case presents an example of the continuities, or long-term changes, in ironworkers' histories, while the second one presents a contrasting example of more rapid change.

Unique and characteristic formal features of the *ikul* knife identify it as a product of Kuba workshops, and serve to distinguish Kuba blacksmiths' work from all others in the region. Its subtle and complex structural geometry, in particular the unusual midrib, is evidence that its designer possessed effective finishing tools and an exceptionally high degree of skill in using them. That many such knives from different workshops were made in precisely the same way testifies to a sustained commitment to advanced standards of excellence on the part of Kuba smiths. After viewing and handling over a hundred examples, one can infer, not unreasonably, that these smiths paid careful attention not only to their own work but to the work of others. One senses a keen ambition on the part of those who invented new products based on the *ikul* as a model. As with many other prestige knives produced in the western rainforest areas of central Africa, the *ikul* blade was embellished by blackening the surface heavily and then selectively polishing it. The geographical distribution of this shared technique, along with the knife's consistently reproduced formal integrity, suggest that the master who invented it lived well before the nineteenth century, a time when engraving came to be in vogue. Its name, *ikul*, offers evidence to support this, as it is derived from Mongo, the language spoken in the ancestral homelands of the Bushoong ruling elite. When we compare *ikul* to the strikingly dissimilar Kuba war knives which were invented outside the kingdom, we begin to see that it typifies the originality of Kuba smiths. It thus becomes more understandable why wearing the *ikul* was considered an essential aspect of Kuba male identity and status.

It makes a strong visual statement about the long-term training investments made by blacksmiths in the kingdom and their ongoing patronage relations with its elites and rulers.

Other smiths, who made fancy hatchets in workshops established around the colonial post at Luluabourg from the 1880s onward, have a very different history, one shaped much more by mercantile values. They assembled the basic form of the hatchet out of separate parts, several lengths of iron fanning out from the handle like fingers, each of their tips welded to a crosswise blade. In selecting the parts and fashioning the openwork structures, smiths were not consistent. Examples of hatchets varied dramatically in the number of iron units, their dimensions, and the scale of the final product. Such variability suggests that the object was a relatively recent invention, and that its formal qualities had not yet been firmly set into standard conventions that were repeated precisely by smiths over time. Comparing it with a similar type of hatchet, though one that was much smaller and more elegantly and carefully worked, highlights the hurried, rather cursory workmanship of the Luluabourg smiths. Moreover, the small hatchet is very rare in museum collections, as it was hardly ever sold to outsiders. Hence, the Luluabourg designs appear to have been new commercial offshoots of an already established and revered masterwork. These newer hatchets were made for and from the market, assembled from plain bar iron currency units that circulated in the Kasai region during the late nineteenth century, and they were called only by a generic name for iron, *kilonda*.[21] As visual statements, they are bombastic and overblown, made by clever blacksmiths in quick response to new marketing opportunities presented by Europeans and their local allies stationed at Luluabourg.

Each of these histories offers a glimpse of particular ironworkers, and generally when, how, and for whom they worked. Until now, ironworkers have seemed hopelessly elusive as historical subjects. They appear only intermittently in what written historical sources exist for central Africa, accounts which are for the most part recent, sparse, scattered, and very uneven in reliability.[22] Other more extensive and reliable evidence of ironworking comes to us by way of archaeological research, but it pertains mostly to smelting operations in the distant past.[23] As mentioned above, oral traditions portray ironworkers only very generally in the form of clichés, while the more detailed evidence available from oral testimonies, colonial reports, and anthropological case studies is inevitably colored to some degree by the particular social and economic conditions of the twentieth century. In short, ironworking was an occupation that left few conventional historical traces, especially for the several centuries preceding colonial rule.

Rich and contemporaneous sources do indeed exist for the history of ironworkers in the precolonial period, although they are not often consulted by historians. They are the tools used and commodities made by ironworkers.

Some of these are still kept in local treasuries, while many others are conserved in museums in Europe, Africa, and North America. Material objects are the most detailed, internally-generated primary sources we have for the history of ironworking, and of blacksmiths in particular. Their eloquence is compounded when other contemporary primary sources are used to corroborate and supplement them. The most reliable of these latter are lexical data on metalworking from western Bantu languages. Thus we have two independent sets of data available that originated in the place and time in question: objects and words that were current in the nineteenth century. These are the major primary sources on which I base this study.

Collections of objects, public or private, are not like archives of written sources and should not be expected to provide the same kinds of evidence. They do not give us sufficiently detailed, precisely dated information about individual workers, their specific careers, and how they and their workshops fared over time. To be sure, they have their limitations, as do documents and private papers, but they also can yield data unavailable anywhere else. In this case, iron products—including ironworking tools, metal implements and weapons, currencies, prestige objects—revealed much about what ironworkers actually did in the recent and more distant past, under conditions that have since disappeared or changed. What is registered in these material sources are historical details about the work as it used to be done: the tools, techniques, levels of skill, and aspects of the labor process, things that one could never expect a smelter or blacksmith living today to know. Thus, as contemporaneous evidence, nineteenth-century metal products also serve to expose biases that occur in oral testimonies and observations of twentieth-century workshops, minimizing the possibility of anachronistic interpretation and other such distortions.

If the material sources are "read" systematically and viewed with an informed eye, handmade objects can reveal the histories of the makers and conventions they followed in their work. Each object is part of a technical and visual language of working, a logic of making, which must be uncovered and understood before it can be interpreted as a historical source. Most manually-produced objects are entry points into much larger groups of historically related objects, morphological sequences extending across space and time,[24] which can reveal chains of transmission of certain constellations of skills and techniques from worker to worker. It is the tracing of continuity and change in workers' conventions within these larger groups of objects that can be most promising for historians.

The important work of the blacksmith and his extensive social relations, which are the subject of this book, began to emerge only when I identified the many overlapping markets and patterns of consumption that existed for iron goods in the region. My distribution maps, which showed the geographical areas where certain groups of objects were made, used, and traded, were, in fact, maps of major market areas. But mapping came at a later stage in the research. I began this project by studying intensively a collection of forty-four metal implements, documenting how, when, and by whom they were collected. Once I had established that they were acquired in the Zaire basin between 1891 and 1905, I then analyzed them, describing each one fully, measuring, weighing, making a scale drawing of it, and sketching its cross-sectional geometry. Slowly, the conventions used in making each object became apparent to me. At the same time, I interviewed and trained with several types of experts in order to hone my skills so that I could better interrogate and read metal objects for the data they might reveal about the tools and techniques used by smiths in giving them form. I worked with an engineer for a blade manufacturer, an experienced foundryman and materials science researcher, and professional blacksmiths. Next I had to gather the assorted objects I had been studying into groups, and it was not at all clear to me which of their features were significant or distinguishing criteria for doing so. I did not rely on apparent similarities among the objects, since it is not necessarily obvious or readily visible even to a trained eye when handcrafted forms are historically related. I therefore turned to cluster analysis as a rigorous technique for dealing with the problem of how to rank order features of objects and form them into groups.[25] It was at this stage that I enlarged my study sample, first surveying and collecting general data on about three thousand metal products in the collections of the Musée Royal de l'Afrique Centrale in Belgium, then composing a total sample of close to two hundred objects.[26] My cluster analyses generated several groups, each one made up of objects that shared certain arbitrary features. Then, to see whether these material sources could provide other historical evidence beyond these technical matters, I set about trying to reconnect them to their social milieux.

I reasoned that if the groups of objects held together in geographical distribution maps, then historical connections between their makers was possible, even likely. I therefore set out to establish the provenance and geographical distribution of each object type, surveying and verifying all known published examples from the nineteenth and early twentieth centuries, lexical data published in written accounts and Bantu language dictionaries, and the collectors' or donors' information for museum objects. In this way I was able to reunite the objects with the probable makers of them in their local contexts. The context was sometimes a language group, sometimes more specifically a town or vicinity, and on occasion, a workshop. I also engaged in comparative analyses to try to postulate morphological histories. Objects must be compared with one another to determine the standardized product from the unusual one, the semifinished from the finished, the common commodity from the prestige object, the utilitarian work from the work of art. Morphological histories show how an object type can move from one of these categories into others over time, changing in certain details according to the needs and desires of consumers. In all, by the end of this research project, I had analyzed additional museum and private collections, surveying and working with over four thousand metal objects made in central Africa during the nineteenth and early twentieth centuries, and creating an archive of primary data.

To broaden and strengthen my evidence, I developed another archive of nineteenth-century primary data: words and vocabularies. The database consists of general and specialized lexical terms for metalworking tools, techniques, and products, and terms related to wealth, prestige, and status. The word lists I compiled are from over 150 western Bantu languages, and were gathered mostly from early published dictionaries and ethnographies, some more reliable than others.[27] I use these data in several ways: to corroborate or critique visual evidence from material objects; to modify inferences drawn from other, weaker evidence such as twentieth-century colonial reports; and to detect biases in oral testimonies and traditions. This database also allowed me to expand my geographical scope from the middle and upper Zaire and Kasai basins to west central Africa as a whole (excluding Kongo and coastal Angola).

Finally, a crucial component of this research project was my field visit to Lopanzo, Equateur, Zaire. There I had the opportunity to observe smelting and smithing first hand, to interview elders and blacksmiths about ironworking in the present and past,[28] and to examine iron goods kept in local treasuries. This helped me to recognize several mistaken assumptions I had had about the working and labor processes, and opened up new lines of inquiry that I had not envisioned before. Even when it was clear to me that what I was learning was relevant only to the present, witnessing the work and workers *in situ* advanced my understanding. Commissioning a skilled blacksmith to make metal prestige objects I had already studied in museums deepened my understanding of what kinds of changes were more possible or likely than others. Regarding production, for example, basic morphologies of most specific standardized products were relatively stable, while surface treatments and other details could be more variable. Although levels of wealth had clearly changed with different economic conditions, consumption patterns revealed a hierarchy of entitlement and preference among metal luxury goods that appeared to be consistent with early twentieth century evidence. Currency forms and their circulation patterns, however, had changed dramatically. These and other findings provided the core of a case study which portrays the social mobility of particular blacksmiths in the late nineteenth and early twentieth centuries.

I describe my sources and methodology as a necessary prelude to offering some explanatory comments on the book's structure and format, and on certain features in the text. Most importantly, the reader will encounter what may seem at first glance to be descriptions of individual objects. They are not. They are, in fact, histories of groups of objects, accounts that I have pieced together to trace what certain ironworkers did over time and for whom. Moreover, they are not based on material sources alone. I built them out of histories of objects and words, supplemented by other available data when it was found to be reliable and appropriate. These data come from many sources, including archaeological reports, oral traditions and testimonies (some seventeenth and eighteenth century, but mostly nineteenth century), nineteenth- and early twentieth-century ethnographies, explorers' and missionary accounts and drawings, colonial reports, archival photographs, case studies by other scholars, and field and laboratory observations. It was these histories, which served as entries into certain smelting and smithing workshops at different times and in different places, that yielded evidence of the broader social and economic aspects of the work and its significance.

The dates I have selected for this study as brackets defining the nineteenth century, ca. 1820s to 1920s, mark general turning points in central African history. Both of these turning points dramatically affected the craft tradition and its workers. The *terminus a quo*, the 1820s and 1830s, was a time when direct trading links to international markets opened up, initiating a period of economic growth during which ironworking flourished. Consolidation of colonial rule during the 1920s signals a new era when the craft tradition declined. My work concerns what ironworkers did in the time between these brackets. Included throughout the text are some of the best-documented and most fully developed examples of the many and interwoven ironworking histories I have been able to trace for the region. And so, as a collection of related but separate histories, it follows a thematic structure rather than a single chronological progression. Moreover, because the primary sources I rely on do not provide precise details of time and place, the chronologies I do discuss are often relative, that is, they are without calendrical dates. Similarly, the overall geographical parameters of the study are general and broadly drawn. What I try to do here is to provide a certain amount of coverage of the region, to shift foci and cast spotlights on various locales to show that what I am discussing at a particular time is not an isolated example but represents a regional trend. My ironworking histories are also selected to convey some sense of the overall range in variability of ironworking skills, markets, and workshops.

Social and economic historians can be unaware or even dismissive of primary sources that are not in written or spoken form. Material objects, for example, tend to be associated with the disciplines of art history or anthropology. That objects play such an important part in this study, however, does not render it something other than history.[29] I borrow as needed certain tools and techniques of analysis from art history, archaeology, and material culture studies; the method, approach, and questions I pose and pursue are unequivocally historical. Above all else, my research is based on primary sources that are contemporaneous with the time period, and I have employed source criticism and followed the rules of evidence in interpreting them. It is my hope, therefore, that readers will recognize this book as the product of specifically historical analysis.

ORGANIZATION OF THE BOOK

I begin in Chapter 2 with a discussion of theoretical models and the importance of empirical evidence for the history of African ironworking. Then I move on to a review of what is known about this craft tradition prior to the 1800s. I end with a brief overview of the historical conditions for the period under study.

Chapters three, four, and five together form a regional survey of ironworking in the nineteenth century. Presenting such a survey with a regionwide perspective serves a dual purpose. First, it provides the reader with a general setting, locating workshops and markets for iron in relation to some of the broader and more relevant historical trends taking place in the region at the time. Second, and most importantly, this broad geographical framework allows me to show how blacksmiths were hardly limited to providing their immediate families and communities with simple tools. Regionwide, their work was divided into three main specialities, each one exhibiting its own particular market patterns: smelting iron, producing metal currencies, and making finished iron products.

Chapter 3 is about smelting, and I focus on furnaces and where they were located. My map shows that smelting expertise was widely but unevenly distributed, the outcome of strategies followed by master smelters. By managing the process of turning ore into iron, they chose where they would work as well as when and to whom they would pass along their esoteric knowledge and skill. Smelting workshops were not clustered around the important centers of trade but instead were often located in remote rural areas that offered security and stability. Thus ironworkers played a part in expanding and integrating the economic zones of the Zaire River basin.

Blacksmiths who could not smelt had to get their iron in some other way, usually in the form of currencies, which are the subject of Chapter 4. While there is some evidence that unrefined iron direct from the furnace was purchased on site or traded, most of the time the metal was refined and hammered into various types of currency units—from standardized spherical or cubic forms to semifinished blades. By mapping the geographical areas in which each was known and recognized, I show how smiths made iron currency forms of one kind or another that circulated all over the region. Moreover, all of the boundaries of these currency zones were elastic and permeable, and it was blacksmiths who did the converting of one currency form into another or into a specific product.

In Chapter 5 I discuss the hierarchy of work in the smithy, from simple sharpening and general repairs to the more challenging tasks involved in making the most important types of finished products. Among the latter were the blacksmith's tools, the most valuable of which, hammers and anvils, were masterworks produced in limited quantities and usually only for newly qualified smiths. Many other tools and weapons were made and marketed for use in hunting, farming, war, and personal protection; still other iron products served to designate social status and leadership positions throughout the region. These standardized products came in many varieties, each revealing stories about the changing social relations between smiths and their clientele in different locales, and how they collaborated over time in developing certain lines of specialized products.

In each of these chapters I show how closely economic specialization was linked to layers of social relationships that ironworkers formed among themselves and with others. I emphasize the fact that smelters and smiths manufactured a rich array of products from the most practical to the fancy, offering many avenues for specialization. Many ironworkers diversified their operations, combining product specializations in different ways to reach a number of markets instead of just one or two. Markets were not simply economic networks but were also social networks, connecting smiths with one another sometimes over hundreds of miles. Iron goods were many and varied, and along with iron goods traveled information about who made them and where. Master ironworkers were well known, and they are the key to tracing change. Ambitious ironworkers trained more than once—they trained and retrained, new skills being sought and acquired from masters across residence and language boundaries. Rather than simply projecting already known social relations onto these workers, I instead show how, through their production and marketing strategies, they initiated new social relations and transformed others.

In Chapters 6 and 7, I present two case studies of ironworking histories in their social and political contexts. The first focuses on ironworking in the Kuba kingdom, a kingdom whose oral traditions contain several blacksmith clichés. These clichés, having to do with smelting, smithing, and great masters and ironworking feats of the past, serve as points of departure for looking at how ironworking was integrated into the kingdom's social and political organization. My evidence shows that ironworkers were independent producers, though smelters and some smiths sent a portion of their iron as tribute to the capital. Highly skilled masters served the administration more directly by turning that iron into insignia that designated nobles and titleholders, some of whom were themselves smiths. Great masterworks of iron that were kept in the king's treasury and sometimes offered as gifts to visitors eventually found their way into museum collections abroad. I conclude therefore that production and distribution of iron was in many ways of central importance to the Kuba kingdom, but that it was encouraged, not controlled by the government.

The second case study is based on my own field visit to the town of Lopanzo in the middle Zaire River area, a town that was founded by blacksmiths. It shows what ironworkers' economic autonomy could mean in a locale where political power was decentralized and population density was relatively low. Here, blacksmiths could move their workshops, establishing what would become new settlements near ore deposits. Although there were other sources of iron circulating in the nineteenth century, being able to smelt and having a dependable source of ore nearby allowed master smelter-smiths to create iron at relatively low cost. If they then invested substantially in titles and women, for which payments were in iron, they earned prestige, "created lineages," and became members of the upper strata of society. After the death of a blacksmith, his own masterworks were kept in treasuries by his heirs, and I witnessed how they then served as production models and also as mnemonic devices to keep a master blacksmith's name and reputation alive.

In Chapter 8, I combine all the foregoing elements into an explanation of ironworkers', especially the blacksmith's, social status and prestige in nineteenth-century west central Africa. Ironworking was potentially profitable work, and offered men a high degree of autonomy. Talent, skill, and entrepreneurship turned a smith into a master, and master blacksmiths used their wealth and economic position as leverage in their social and political relations. They became "big men" with large households in a town along the middle Zaire, and they became aristocrats and titled officials in the Kuba kingdom. But I argue that this economic base, important though it is, supplies only one important facet of the story. As both my regional survey and case studies show, the uses and significances of iron went well beyond its material and utilitarian purposes. The finest products of blacksmiths generated more than wealth alone. Successful smelts were great performances and meeting places where the work and its masters were made famous, while smithies were the centers where gossip and news was collected and exchanged. Masterworks of blacksmithing—hammers, anvils, and ceremonial objects—represented both the skills of the maker and the skills of the owner, and communicated visually various kinds of messages about status, position, legitimacy, and history.

Chapter 8 also includes a discussion of the roles played by ideology and culture in the histories of ironworking. By successfully excluding women from the occupation, smelters and blacksmiths created a male identity of high status and prestige. They also claimed to keep the occupation within the family or "lineage," through mostly father-to-son training. Thus if one accepts the stated norms, one might infer that lineage ideology or culture acted to constrain the spread of ironworking techniques—that culture in effect hampered historical change. But not only were there admitted exceptions to the normative pattern of training, there was also evidence of how culture could act as a catalyst and vehicle for historical change. The prohibitions masters deployed to maintain control over their skills helped make iron valuable and served as well to promote their own reputations. So too, did the visual needs of their oral societies, where iron insignia and prestige objects were a locus of innovation. Blacksmiths knew about each other, competed with one another, and masters might be open to negotiation to teach special or new techniques to certain selected men, if the bargaining terms were right. Smelters and smiths from different ethnic groups in the region shared particular types of equipment, tools, techniques, and terminologies, which would not be the case if monopoly control had been complete. There was both continuity and change in ironworking, and cultural factors played a part in each.

I conclude by stressing that the mystique of the ironworker is a myth of his own making. There were undoubtedly ironworkers who failed, who were poor, and who were not remembered. And there were other craftsworkers whose products were economically and socially important, but not in as many ways as the ironworkers' were. The myth of the blacksmith was made by the masters themselves who created, maintained, and revised the technology and the occupation, and who orchestrated the processes of smelting and smithing and the rituals that accompanied them. All of this complexity comes unraveled in the twentieth century, an untold story which I sketch out in the Epilogue. What remains today is pride. The wealth and social prestige of ironworkers were marks of their success, success which they themselves aspired to and used to their own advantage, and which others in their societies acknowledged by respecting and remembering them.

NOTES

[1] See Eugenia Herbert, *Iron, Gender, and Power. Rituals of Transformation in African Societies* (Bloomington: Indiana University Press, 1993), especially pp. 78–97.

[2] See David Birmingham and Phyllis Martin, eds., *History of Central Africa* (London: Longman, 1983); Joseph Miller, *Kings and Kinsmen: Early Mbundu States in Angola* (Oxford: Clarendon, 1976); Anne Hilton, *The Kingdom of Kongo* (Oxford: Clarendon, 1985); John Thornton, *The Kingdom of Kongo: Civil War and Transition, 1641–1718* (Madison: University of Wisconsin Press, 1983); Idem, *Africa and Africans in the Making of the Atlantic World, 1400–1680* (Cambridge: Cambridge University Press, 1992); Thomas Reefe, *The Rainbow and the Kings: History of the Luba Empire to 1891* (Los Angeles: University of California Press, 1981); Jan Vansina, *The Tio Kingdom of the Middle Congo, 1880–1892* (London: Oxford University Press, 1973); Idem, *Children of Woot* (Madison: University of Wisconsin Press, 1978); Idem, *Paths in the Rainforests* (Madison: University of Wisconsin Press, 1990); Georges Dupré, *Un Ordre et sa Destruction. Anthropologie des Nzabi* (Paris: ORSTOM, 1982); and Marie-Claude Dupré and Bruno Pinçon, *Métallurgie et Politique en Afrique Centrale* (Paris: Karthala, 1997).

[3] R.F. Tylecote, *A History of Metallurgy* (London: The Metals Society, 1976). For the most lucid discussion of the complexities of bloomery smelting, see David J. Killick,

"Technology in its Social Setting: Bloomery Iron-Smelting at Kasungu, Malawi, 1860–1940" (Ph.D. diss., Yale University, 1990).

[4] See Tal Tamari, "The Development of Caste Systems in West Africa," *Journal of African History* [hereafter *JAH*] 32 (1991); Adria LaViolette, "An Archaeological Ethnography of Blacksmiths, Potters, and Masons in Jenne, Mali (West Africa)" (Ph.D. diss., Washington University, 1987); David C. Conrad and Barbara E. Frank, eds., *Status and Identity in West Africa* (Bloomington: Indiana University Press, 1995); and James Quirin, *The Evolution of the Ethiopian Jews* (Philadelphia: University of Pennsylvania Press, 1992).

[5] This difference was first pointed out by Pierre de Maret, in "Ceux qui Jouent avec le Feu: la Place du Forgeron en Afrique Centrale," *Africa* 50, 3 (1980). My evidence corroborates his.

[6] The archaeological studies in the vicinity of Jenne are an exception. See, for example, Susan Keech McIntosh and Roderick J. McIntosh, "From Stone to Metal: New Perspectives on the Later Prehistory of West Africa," *Journal of World Prehistory* 2, 1 (1988).

[7] See Patrick R. McNaughton, *Secret Sculptures of Komo* (Philadelphia: Institute for the Study of Human Issues, Inc., 1979), especially pp. 10–23; and Idem, *The Mande Blacksmiths* (Bloomington: Indiana University Press, 1988).

[8] George E. Brooks, *Landlords and Strangers: Ecology, Society, and Trade in Western Africa, 1000–1630* (Boulder: Westview Press, 1993), pp. 46, 74. He relied on McNaughton in constructing this argument.

[9] See, for example, Walter Cline, *Mining and Metallurgy in Negro Africa* (Menasha, Wisconsin: Banta, 1937); and S. Terry Childs and David Killick, "Indigenous African Metallurgy: Nature and Culture," *Annual Review of Anthropology* 22 (1993).

[10] See Ralph Austen and Daniel Headrick, "The Role of Technology in the African Past," *African Studies Review* 26, 3/4 (September/December 1983); the debate in *African Economic History* 19 (1990-1); and Thornton, *Africa and Africans*, pp. 45–48. A particularly heated debate arose over the claim that African smelters had discovered a unique method for producing high temperatures in their furnaces. See Peter Schmidt and Donald Avery, "Complex Iron Smelting and Prehistoric Culture in Tanzania," *Science* 201, 4361 (22 September 1978); J. E. Rehder, "Use of Preheated Air in Primitive Furnaces: Comment on Views of Avery and Schmidt," *Journal of Field Archaeology* 13 (1986); Manfred Eggert, "Katuruka und Kemondo: zur Komplexität der frühen Eisentechnik in Afrika," *Beiträge zur allgemeinen und vergleichenden Archäologie* 7 (1985); Idem, "On the Alleged Complexity of Early and Recent Iron Smelting in Africa: Further Comments on the Preheating Hypothesis," *Journal of Field Archaeology* 14 (1987); and David Killick, "On Claims for 'Advanced' Ironworking Technology in Precolonial Africa," in *The Culture and Technology of African Iron Production*, ed. Peter Schmidt (Gainesville: University Press of Florida, 1996). See also D. Avery and P. Schmidt, "The Use of Preheated Air in Ancient and Recent African Iron Smelting Furnaces: A Reply to Rehder," *Journal of Field Archaeology* 13 (1986). Most, but not all, of these articles are reprinted in the volume edited by Schmidt, *Culture and Technology*. For a more rigorous comparative study, see Robert Gordon and David Killick, "Adaptation of Technology to Culture and Environment: Bloomery Iron Smelting in America and Africa," *Technology and Culture* 34, 2 (1993).

[11] In 1899, Grenfell, who was one of the more reliable observers of ironworking, witnessed smiths working in the Aruwimi River basin. The iron he described was very soft and tough, and was hardened by hammering. George Hawker, *The Life of George*

Grenfell (New York: Fleming Revell Co., 1909), pp. 446–47. It is often assumed that quenching is the best way to harden an iron cutting edge; however, there are clear advantages to hammer hardening. See J. E. Rehder, "Iron versus Bronze for Edge Tools and Weapons: A Metallurgical View," *Journal of Metals* (August 1992).

[12] Ironworking was presumed to have been one reason why Bantu speakers prevailed, though recent empirical studies in linguistics and archaeology have shown that proto-Bantu speakers were not ironworkers at all, and that the change from using stone tools to iron ones did not result in immediate and dramatic societal change. Nevertheless, for over a century, the study of Bantu languages and the study of early ironworking have been much entangled and difficult to separate. For critiques, see Jan Vansina, "Bantu in the Crystal Ball," Parts I and II, *History in Africa* 6 (1979) and 7 (1980); and Manfred Eggert, "Historical Linguistics and Prehistoric Archaeology: Trend and Pattern in Early Iron Age Research of sub-Saharan Africa," *Beiträge zur allgemeinen und vergleichenden Archäologie* 3 (1981). For a discussion of the linguistic evidence, see Pierre de Maret and F. Nsuka, "History of Bantu Metallurgy: Some Linguistic Aspects," *History in Africa* 4 (1977); P. de Maret and G. Thiry, "How Old is the Iron Age in Central Africa?," in *Culture and Technology;* and Jan Vansina, "New Linguistic Evidence and 'The Bantu Expansion'," *JAH* 36 (1995). For an updated view of "Bantu expansion," see Vansina, *Paths*.

[13] Here I am referring especially to Mircea Eliade's 1956 book *Forgerons et Alchimistes (The Forge and the Crucible)*, which became a classic in the field of comparative religion and mythology. In it, he gathered together for the first time an extraordinary array of myths and religious rites from all over the world that, in his view, conferred upon metalworkers an almost sacred status. He admitted that he was concerned almost exclusively with the ritualist behavior of the metalworker, not with the general history of metallurgy or the spread of metallurgical techniques. So we are left with the impression that the special status of the blacksmith is the product of a spiritual motivation and an ideology based on the sacred nature of materials and technological processes. Similar circular arguments can be seen in structural anthropological interpretations of ironworking, the most prominent being Luc de Heusch, "Le Symbolisme du Forgeron en Afrique," *Reflets du Monde* 10 (1956). Although these sources are dated, the stereotypes of ironworkers that they present are often encountered today, even among scholars. See Colleen Kriger, "A critical look at Mircea Eliade and the myth of the mystical blacksmith" (paper presented at the African Studies Association annual meeting, Columbus, Ohio, 13–16 November 1997.

[14] For example, see Hilton, *The Kingdom of Kongo*; Miller, *Kings and Kinsmen*; Vansina, *Children of Woot*; and Reefe, *The Rainbow and the Kings*. For the most thorough historical investigation of metalworking and the legitimation of political authority, see Zdenka Volavka, *Crown and Ritual: Royal Insignia of Ngoyo* (Toronto: University of Toronto Press, forthcoming).

[15] See Reefe, *The Rainbow and the Kings*, pp. 82–85. Reefe takes the references to early founders' expertise in ironworking literally and combines them with evidence of iron ore and salt deposits having been exploited in at least the recent past (19th century) in the areas where early Luba dynasties ruled. He fails to note, however, that the most famous early leader was characterized as a smith, not a smelter, and that he reportedly summoned a well-known smelter to come and disseminate the knowledge and expertise of smelting among many Luba-speaking groups and also to communities of Songye, their neighbors to the north. He also neglected to consult geological sources which comment on how widespread and frequent iron ore deposits are in the Zaire basin region, espe-

cially in Shaba (see Chapter 2). Thus his explanation of early state formation strains the evidence, at least with regard to iron.

[16] See P. de Maret and F. Nsuka, "History of Bantu Metallurgy."

[17] An especially voluminous literature has grown around the study of slavery, women, peasants, and wage laborers. See, for example, Joseph Miller, *Slavery, a Worldwide Bibliography, 1900–1982* (White Plains, NY: Kraus International, 1985); Idem, *Slavery and Slaving in World History: A Bibliography, 1900–1991* (Millwood, NY: Kraus International, 1993); and bibliographies in Catherine Coquery-Vidrovich, *African Women: A Modern History* (Boulder: Westview Press, 1997); Jane Parpart, "The Labor Aristocracy Debate in Africa," *African Economic History* 13 (1984); and Bill Freund, *The African Worker* (Cambridge: Cambridge University Press, 1988). See also Catherine Coquery-Vidrovich and Paul Lovejoy, eds., *The Workers of African Trade* (Beverly Hills, CA: Sage, 1985).

[18] For example, Frederick Cooper, *From Slaves to Squatters* (New Haven: Yale University Press, 1980); Luise White, *The Comforts of Home: Prostitution in Colonial Nairobi* (Chicago: University of Chicago Press, 1990); Elias Mandala, *Work and Control in a Peasant Economy* (Madison: University of Wisconsin Press, 1990); and, more recently, Sara Berry, *No Condition is Permanent* (Madison: University of Wisconsin Press, 1993); Keletso Atkins, *The Moon is Dead! Give us our Money!* (Portsmouth, NH: Heinemann, 1993); Patrick Harries, *Work, Culture, and Identity* (Portsmouth, NH: Heinemann, 1994); Allen Isaacman, *Cotton is the Mother of Poverty* (Portsmouth, NH: Heinemann, 1996); Frederick Cooper, *Decolonization and African Society: The Labor Question in French and British Africa* (Cambridge: Cambridge University Press, 1996).

[19] Most detailed research on artisanal work in Africa has been carried out by anthropologists and archaeologists. Historical studies (besides my own) include works by Marion Johnson, Ann O'Hear, Philip Shea, Richard Roberts, Judith Byfield, Carolyn Keyes Adenaike, and others. These are written mainly from economic perspectives, and some of them are still unpublished. For initial entry into the literature involving preindustrial artisans and the so-called "labor aristocracy" of industrialized working classes in Britain, continental Europe, and the United States, see E.P. Thompson, *The Making of the English Working Class* (New York: Vintage, 1963); Eric J. Hobsbawm, *Labouring Men. Studies in the History of Labour* (London: Weidenfield and Nicolson, 1964), especially Ch. 15; Idem, *Workers: Worlds of Labor* (New York: Pantheon Books, 1984), Chs. 12, 13, and 14; Joan Wallach Scott, *The Glassworkers of Carmaux* (Cambridge: Harvard University Press, 1974); Ronald Schultz, "The Small-Producer Tradition and the Moral Origins of Artisan Radicalism in Philadelphia, 1720–1810," *Past and Present* 127 (May, 1990); and John Smail, "Manufacturer or Artisan? The Relationship between Economic and Cultural Change in the Early Stages of the Eighteenth-century Industrialization," *Journal of Social History* 25, 4 (1992). The thesis that artisans and a skilled "aristocracy" of labor were somehow inherently less radical politically than other, less-skilled workers has been disproven repeatedly in local empirical studies.

[20] In this study I de-emphasize ethnicity, but I do not mean to imply that it did not exist in precolonial times. The important research on the creation of "tribes" during the colonial era has sometimes been taken too far. See, for example, the exchange between Alex de Waal and Jean-Luc Vellut, the former discussing colonialists' creation of distinct ethnic groups in Rwanda, and the latter's deeply informed response, demonstrating the pitfalls of neglecting and over-simplifying the precolonial African past. *The Times Literary Supplement* (1 July and 15 July 1994).

[21] The hatchet itself was called *kilonda* in Songye (L23), *cilonda* in Luba Kasai (L31a). *Kilonda* is a word meaning "iron" in Luba Shaba (L33) and Hemba (L34), the Songye form being *bilonda.*

[22] With the exception of written sources about Kongo and Angola, although even these say very little about ironworking and ironworkers before the 1760s. Jan Vansina, personal communication, 29 November 1996.

[23] See Duncan Miller and Nikolaas Van Der Merwe, "Early Metalworking in Sub-Saharan Africa: A Review of Recent Research," *JAH* 35 (1994).

[24] I am familiar with an extensive art historical, anthropological, and material culture literature, but much of it lacks rigorous source criticism. My most useful guide in working with objects as historical evidence has been the masterpiece of methodology by George Kubler, *The Shape of Time* (New Haven: Yale University Press, 1962).

[25] See my discussion of them, and their practical application for historical research, in Colleen E. Kriger, "Museum Collections as Sources for African History," *History in Africa* 23 (1996).

[26] For a more complete discussion of the sample and how it was composed, see Kriger, "Museum Collections," p. 138.

[27] I am aware of the many problems surrounding how languages were selected and distorted by orthography. Moreover these dictionaries and ethnographies often lack the technical terminologies I would have liked to have had at my disposal. In some cases, however, a wealth of evidence could be gleaned from these sources, particularly when juxtaposed with the visual evidence.

[28] Interviews with blacksmiths were informal, daylong affairs during the work in progress at the smithy, and were conducted in French, through interpreters. I gleaned much of my information by observing the work and asking very precise questions about what was being done and for what purpose. In my experience interviewing artists and artisans in Europe, Africa, North America, and Japan, I have found that what matters most is not the verbal language one speaks but how well one understands the work and work-related issues.

[29] More than once I have been labeled an art historian, though my degree is in history and my sources, methodologies, and questions extend well beyond conventional art history as practiced in the English-speaking world. The material objects I consult as sources range from currency units to utilitarian implements to ceremonial objects, all for the purpose of gleaning from them whatever it is they can tell us about ironworkers, their work, and their social and economic networks in the central African past. The range of types of objects is closer to what is called "material culture" than it is to a conventional art historical selection. But this is not simply a narrow study in "material culture," nor is it precisely a study of *mentalités*. This latter approach, developed so fruitfully by the Annales school, is usually associated with history "from below." It should be obvious to the reader that I share much in common with those Annales historians who have engaged in research on social history, using unconventional primary sources. Nevertheless, the final result of my research is a history of an elite, though an elite that until now has been "invisible" to historians and that was displaced by new elites in the twentieth century.

2

Iron Production in Central Africa: Building a Craft Tradition, ca. 600 B.C.E. to 1920

This chapter consists of three main sections, a tripartite foundation for the chapters which follow. In the first section, I address problems having to do with the limitations of our presumed knowledge about metallurgy in world history, and show why technological details can matter so much when studying a group of artisans. I describe several examples of how specific, empirical evidence from ironworking practice in Africa challenges commonly held assumptions about the technology, and, by extension, the workers.

In the next section, I provide a brief survey of archaeological evidence for smelting and smithing in central Africa, from the earliest times up to the period 1100 to 1400. In doing so, I place my study of nineteenth-century ironworkers in a much broader temporal context, demonstrating that their history was long in the making. From the beginnings of their craft tradition over two and half millennia ago, central African smelters and smiths had a hand in generating the trends toward greater wealth and social complexity.

The chapter ends with an overview of the historical setting, first calling attention to the lacunae in our knowledge of ironworking in central Africa after 1100, and then outlining the general political, social, and economic changes that were taking place there during the nineteenth century. I conclude by suggesting that ironworking at this time represents the craft tradition at its height.

CENTRAL AFRICAN IRON SMELTING: EVIDENCE AND MODELS

Making iron metal out of iron ores has been a challenge in the past as it continues to be in the present. Nevertheless, it is common to come across references to nonindustrial metallurgical processes that lump them together and describe them as "simple." Some scholars are among the many who make such errors. Forbes Munro, for example, wrote that because the extraction of metals in precolonial Africa was such a "simple" technological process, economic historians of Africa could analyze the trade of unworked metals together with other "foraged or collected" commodities such as elephant tusks and kola nuts. The cavalier nature of this assertion fails to conceal the author's fundamental mistakes in conflating altogether different types of resources, and confusing mineral ores with metals.[1]

Going back almost a century, a similarly cavalier tone marked comments by Belgian promoters of the Congo Free State who, in explaining the prevalence and high degree of specialization in indigenous ironworking there, attributed it to the widespread availability of ores. Characteristically boasting about the natural resources of King Leopold's colonial enterprise, officials wrote with confidence that the Zaire basin contained an abundance of ferruginous minerals, and that the most common form, laterite,[2] was easy to smelt. Thus as ironworking became a theme used in promoting the colonial cause, it was depicted in a highly selective way. Propagandists praised the advanced state of ironworking in the region, which they portrayed as the inevitable consequence of a wealth of laterite ore deposits.[3]

Both of these descriptions miss the point entirely, for they emphasize the raw material or natural resource, metals or mineral ores, while disregarding the workers and the work. Each description conveniently avoids the most important historical dimensions of ironworking: the human activity and human resources essential to directing the technological processes used in producing metal out of ore.

Iron metal is produced by people, not by nature. It is not simply gathered or collected, and the presence of a mineral ore deposit near a human settlement does not guarantee that it will be exploited or that smelting technology will develop. One must be taught to recognize which rocks are ores, for example, which only sets the stage for the rest of the technical and labor processes, all required if iron is to be successfully won from ore.

Such cursory dismissals of nonindustrial technologies can easily occur, especially when they are based on implicit (and often inappropriate) comparisons with the technologies of industrialized societies. When assessing evidence of African iron smelting, it is crucial to keep in mind that our scientific knowledge about metallurgy is not at all comprehensive. Much is known, for example, about the large-scale blast furnaces that began to be used in Europe after the twelfth century, for it was that technology that became the subject of much experimentation and scientific study. Much less is known about other smelting technologies. Small-scale bloomery smelting, which continued in Europe alongside the blast furnace, became economically uncompetitive by the mid-nineteenth century. As a technology apparently in decline, it was not a major subject of scientific study, and is still poorly understood. And so, although the furnaces and the process itself are sometimes described as "simple" or "primitive," attempts to replicate bloomery smelting by experiment in the laboratory often prove to be unsuccessful.[4] In contrast, smelters in Africa, who did not have the theoretical scientific knowledge of how iron was reduced in the bloomery process, were able to produce bloomery iron well into the twentieth century. Replications or reconstructions of smelting procedures on-site in Africa are of interest today for several reasons: for understanding African history; for better understanding how the bloomery process works; and for establishing how it differs from industrial smelting technology.[5]

Clearly, producing iron out of ore in small-scale furnaces is not at all "simple." Nor are identifying, mining, and preparing ore for a smelt simple matters either. To start with, varieties of ores occur in different deposit contexts and with different physical characteristics and levels of iron content, all of which have effects on the smelting process. Access to ore is certainly an important factor, but so, too, is the specific type of ore and how it is treated before and during the smelt. With this in mind, it is worth returning to the questions raised earlier about the extent of mineral resources in the Zaire basin region: what were the ore resources, where were they located, and most importantly, what did ironworkers do with them?

When promoters of the Congo Free State regime hailed the abundance of iron ores in the colony, they were not entirely wrong, but they were not exactly right, either. Among other things, they were making the common mistake of viewing the raw material through modern industrial lenses. For while it is true that the Zaire basin region is rich in iron resources, specific knowledge of its mineral wealth is quite incomplete and skewed. Thorough geological surveys have not been carried out uniformly over the entire region, and what geological information exists comes from a biased colonial and industrial perspective aimed at exploiting African mineral ores to feed the modern blast furnaces of Europe. In the most comprehensive and detailed source for the geology of the region, it was readily admitted that laterite ores, used by most local smelters, were not of interest for industrial purposes,[6] and therefore they were not systematically mapped in official geological surveys. Trained geologists prospected the colony not for the purpose of better understanding current or past mining and exploitation, but to assess the likelihood of generating future revenues in international markets.

In those few instances where colonial geologists wrote about local African iron smelting, they included speculations misleadingly based on industrial practice. The geologist Jules Cornet, one of the earliest geological surveyors and probably the one most sympathetic to and observant of local ironworking processes, stated that to his knowledge, central African smelters only rarely used magnetite and hematite ores.[7] He went on to suggest that these types of iron ore, being very compact, would be difficult to reduce and would probably require the addition of fluxes,[8] thus offering an explanation of why they apparently had not been exploited. His reasoning was flawed, however, and field demonstrations by central African smelters showed researchers how the problems of ore density could be easily avoided.[9] Cornet had incorrectly assumed that African miners would separate the rich ore out from the layers of quartz, which is what industrially oriented miners would do, and so he had mistakenly exaggerated the difficulty of smelting with it.

Cornet also registered comments in his reports about the type of iron ore found most frequently in the Zaire basin, and here again, we discover that empirical evidence challenges his speculations. Laterite ores, which occur primarily in shallow or surface deposits, contain a lot of gangue that acts as a natural flux, and these were the types of ores local smelters used most often in their furnaces. In this case, the presence of fluxes led Cornet and others to conclude that the ore would be easy to smelt. But evidence from actual smelting operations shows otherwise, with some laterite ores presenting their own special problems to ironworkers. Chemical analyses of specific samples of laterite reveal how variable it could be in both iron content and gangue content, which made the selection and sorting of ores especially important.[10] Moreover, certain smelters were apparently able to win metal from even the very difficult, lowest grade ores. In the Ruki River area along the middle Zaire, for example, an archaeological team collected slags and ores, had them chemically analyzed, and came up with some surprising results. Indications are that in this case iron reduction was accomplished using ores so low in iron that successful smelts should not have been possible![11] Exploring just this one aspect of smelting, the types of ore, demonstrates above all that African bloomery smelters adapted their work to a wide variety of conditions and a variety of ore resources, and along the way developed kinds of expertise that were unanticipated even by experienced geologists and metallurgists.

Similarly surprising results come from systematic analyses of smelting furnace designs. The sheer variety of central African bloomery furnaces as well as their specific features call into question some of the received wisdom about how to classify metallurgical processes and define temporal periods in metallurgical development. Conventional typologies of smelting furnace design were based on known cases of early iron smelting in Europe, with bowl furnaces ascribed to the Early Iron Age, and the Roman-style shaft furnace characterizing the Late Iron Age. The later design was considered more "developed," for it included built-up walls and a slag-tapping feature that allowed slag to run out of the furnace while it operated. Moreover, in this later phase, smelters worked with larger quantities of ore, and the working process was more differentiated—roasting the ore preceded smelting, and after smelting the blooms of iron were worked up in a special forge hearth.[12] In short, the European typology follows an evolutionary progression, with general technological and economic advancements identified by greater differentiation in the process, linked in turn to greater efficiency of production and high volume yields.

But the problem is that much of the African evidence does not fit neatly with these criteria. For example, the furnace I observed being used in Lopanzo, Equateur, Zaire, was a shallow bowl, but it also included a slag-tapping tunnel, and the ore was roasted before smelting.[13] Similar ones existed elsewhere, such as in northeastern Zaire and along the middle Zaire in the area around Mbandaka.[14] Their designs combine the relatively small-scale, small-yield features of so-called Early Iron Age furnaces along with other features that suggest the resourceful use of materials and differentiation of tasks associated with intensified production of the Late Iron Age. These and other examples from Africa defy the categories commonly used to classify furnaces.[15] As a result, they call into question certain entrenched beliefs about developments and change in smelting technology, beliefs based on models that originally considered only a small slice of historical experience.

Also following an evolutionary progression is the pyrotechnical model, which has it that metallurgy developed further as metalworkers were able to achieve higher and higher furnace temperatures. This model seeks to explain the sequence Stone Age to Bronze Age to Iron Age as a function of the melting point temperatures of certain metals. Lead first, then gold and copper, with relatively low melting points, were found to have been worked first by ancient metallurgists, before they worked iron. Presumably, this was because the ancients did not have the technical know-how to produce furnaces that could achieve the higher temperatures needed to smelt iron.[16] They were, according to the model, able to develop the ability to achieve higher furnace temperatures only after sufficient experience and experimentation with smelting ores such as copper at lower temperatures. Such is the progression which took place in the ancient Middle East and Mediterranean regions, and it is firmly entrenched in archaeological thinking.[17]

But as knowledge about the bloomery process grows, the pyrotechnical model seems less convincing. Laboratory experiments conducted by metallurgists show that the pyrotechnical model was based on incomplete knowledge of the capabilities of bloomery furnaces. Contrary to what was believed, generating high temperatures in such furnaces was not that difficult. Small-scale bowl furnaces and shaft furnaces were capable of producing temperatures needed for smelting iron, that is, about 1200–1300°C.[18] More importantly, careful study of the bloomery process reveals that the metallurgical goal of smelting iron had not been fully understood, and that it was not simply to reach a very high temperature. Instead, the task of iron smelters was to keep the temperature relatively *controlled*—within that 1200–1300°C range—and to keep the reducing atmosphere of the furnace from creating cast iron.[19] In other words, air supply rate was the most important factor in the bloomery process, with the right air supply rate allowing smelters to control temperature and create the right reducing conditions inside the furnace. An image of smelting as a high temperature, high energy-consuming endeavor corresponds well with the evidence for blast furnaces but not for small-scale bloomery ones. Ergonomic calculations show that there is nothing inherently labor-intensive about manually produced air supply: one man operating a bellows can generate the conditions necessary for smelting iron.[20]

Despite these serious questions, residues of evolutionary and diffusionist thinking still shape the way metallurgical research questions are framed and posed. What remains to be seen is whether or not the rise of iron metallurgy in Africa followed patterns similar to those in other parts of the world. So far, the earliest dates for and the locations of ancient smelting workshops do not show any clearly discernible historical pattern. Some archaeologists are working to find out whether there is evidence of a prior copper technology out of which iron smelting could have developed. Other scholars are assuming that without the prior copperworking foundation, the knowledge of iron metallurgy in Africa must have been introduced from somewhere else, the most prominent candidates being the Phoenicians via Carthage. Still others are suggesting a combination of internal developments and external influences, with smelting expertise then transferred out from multiple centers. What is certain, though, is that as more data on African iron smelting come in, they should not be adjusted or edited to fit classification schemes or hypothetical models that may no longer be entirely satisfactory.

EARLIEST EVIDENCE OF CENTRAL AFRICAN IRONWORKING TO CA. 1400

It was once thought that before Bantu speakers came to inhabit much of central and southern Africa, they already knew how to work iron. Indeed, according to some earlier Bantu expansion models, it was their iron-using and iron-making capabilities that supposedly gave them their competitive advantage over what populations already inhabited the regions they began to settle. Support for this belief came mainly from early linguistic research comparing Bantu languages. It is a belief that lingers on, despite valid critiques of the methodology used in that research and the findings of recent, more rigorous linguistic analyses.[21] Also lingering on is an assumption that ironworking brings unmistakably clear and overwhelming advantages to a social group, allowing that group to easily conquer or displace others. Archaeological findings in West Africa and elsewhere suggest instead that shifts from stone using to iron using did not usher in immediate or major social and economic impacts of this kind.[22]

Linguistic evidence from the Bantu languages remains inconclusive on the subject of where and how early developments took place in ironworking. It does suggest, however, that it is very doubtful that proto-Bantu speakers knew how to work iron.[23] Instead, smithing and smelting were practiced in different societies of Bantu speakers during their dispersals into and within central and southern Africa, not before. But research in historical linguistics tells very little of this story as yet.[24] Part of the problem comes from incomplete data, the word lists and dictionaries that have been compiled not usually having the level of detail that would include very precise and specialized vocabulary terms such as "smelter," or "smelting furnace." So when a word for "blacksmith" is known, it is problematic; it is not clear whether that word means both smith and smelter, if another word for smelter has been overlooked, or if indeed only blacksmithing was known by most speakers of that language. Some tantalizing glimpses and impressions do emerge, however. For example, a number of different words for "blacksmith" were coined in Bantu languages, clusters of them formed around each of several Bantu roots, which suggests multiple local origins for knowledge of the practice of smithing. Among these clusters are two major ones, each derived from basic verb stems. These verbs refer to smithing activities in the forge, one meaning, "to hit," the other "to blow," actions of hammering iron and working the bellows.

Aside from brief linguistic glimpses such as this one, there is relatively little evidence of early blacksmithing that has survived. Though the archaeological data for both ironworking processes, smithing and smelting, are still very limited, far fewer exist for smithing. Unlike smelting furnaces and their slag heaps, the forge does not usually leave prominent marks on the landscape that can be surveyed by archaeologists. Sometimes forge equipment such as clay tuyères can be recovered from excavations, but most often these finds are fortuitous rather than the result of deliberate survey work. Moreover, because iron metal deteriorates in rainforest soils, there are few surviving smithed objects dating from the times for which we have early smelting dates, and when they do exist they are often in very bad condition. When archaeologists' reports mention them at all it is usually only briefly. Nevertheless, the presence of smelting operations presupposes smithing expertise and a market for iron. And so it is the archaeological dates for iron smelting that will have to suffice as baseline temporal indicators for some of the earliest centers for blacksmithing in the region.

Even though the origins of ironworking in central Africa are still incompletely and poorly known, some very general patterns of change can be traced from what archaeological evidence has so far been uncovered for the Zaire basin. It is not an easy matter, however, to sift out what is direct evidence for iron smithing and smelting from the large and growing archaeological database on early settlement. Some archaeologists use the term Early Iron Age to refer to food producing or sedentary populations, while others use it to refer to iron using or iron producing populations. They may identify a shift to an Early Iron Age in Africa on the basis of changes in tools or changes in the morphology and manufacture of ceramics. Therefore it can be instructive to go back to the primary archaeological data to find only those dates that are linked directly to the presence of iron furnaces and iron products. But even then, one must evaluate those dates carefully, and not simply accept at face value all those that are reported, especially the ones associated with the earliest furnace and slag material.[25] Some very early dates reported for smelting in central Africa must be considered unacceptable until more corroborative data come in.

Archaeologists have focused their research on two main geographical areas. The longest archaeological research program for uncovering ancient evidence of smelting in the Zaire basin region focused on sites in the interlacustrine area to the east. Several groups of archaeologists have returned there periodically, accumulating a large body of data including direct evidence of iron smelting taken from furnace remains. In the literature, then, one encounters an emphasis on "the east" or "the lakes." This may be historically justifiable or may instead be merely the result of the research program having been so extensive and well known. In more recent years, a considerable amount of archaeological work has been carried out in west central Africa, in areas north of the lower Zaire, which brings some balance to the picture. Beginning in the early 1980s, two teams of researchers began surveys and excavations in Gabon, centering on lithic industries and early settlement, but also looking for evidence of early metalworking.[26]

These two areas have been identified as the major centers known for the earliest development of iron smelting in the Zaire basin region, one in the east around the lakes, and one in the west, in Gabon. As yet there is no convincing evidence to suggest any direct historical link between these two areas, and no furnaces in other locations have yet been excavated that are as early as these. In general, individual dates are earlier for the lakes, but Gabon also shows an earlier cluster. In the lakes area there are several dates during the period ca. 2500 to 2300 bp, and a more solid cluster of eight dates for Butare (Rwanda) ca. 1700 bp. For Gabon, there are several dates for the period ca. 2200 to 2100 bp, and a more solid cluster of eight dates during the period ca. 2100 to 2000 bp.[27] In both centers the furnaces were the shaft-bowl type but they were constructed quite differently, the lakes version having had its superstructure built up of coiled clay while in Gabon it was made of clay over a wooden framework.[28] Claims that the two centers were historically related thus strain the evidence.[29]

From at least these two early centers ironworkers transferred technologies of bloomery smelting to other blacksmiths in contiguous geographical areas. By the last centuries B.C.E. and first centuries C.E., iron metallurgy was known and being practiced in other parts of west central, east central, and southern Africa. The direct archaeological evidence is still insufficient for discerning any clear patterns in what must have been a complex process of borrowing and adaptation, as blacksmiths proliferated and sought reliable sources of iron metal, and as smelters worked to control conditions in the furnace.[30]

On the eastern side of the Zaire basin, other smelting areas have been identified to the south of the lakes, though one should resist placing them into a single pattern of southward technology transfer. At Kalambo, near the southern tip of Lake Tanganyika, sites yielded iron slag and furnace tuyère fragments at occupation levels dating to the fourth century C.E.[31] Still further south, there were periods of intensive iron smelting in several areas of what is now Zambia.[32] Also in this vicinity, copperworking was carried out in an area straddling the Zaire/Zambia borderlands, now known as the Copperbelt. Archaeologists have been interested above all in the cast copper currency ingots that were produced there, as well as the relation of copperworking to ironworking. Both metals were exploited at about the same time.[33] Copper smelting of ores has been dated at the Kansanshi copper mine to between the fifth and seventh centuries C.E., and this period was followed by two more of similar intensity until the fourteenth century. Then, from the fourteenth to the seventeenth century, there was a marked increase in copperworking.[34]

The most complete direct evidence so far for smithing in the early history of the Zaire basin region comes from metalwork that had been preserved in special protected conditions. Several teams of archaeologists uncovered rich stores of grave goods buried in ceramic containers in the Upemba depression, an area of lakes and swamps in the southeastern part of Zaire. Although it lies along the outer reaches of the Zaire basin, it is an area of importance not only because of the unusual wealth of materials found in the graves, but also because of the long time period over which the burials took place. Using systematic analyses of these metal objects, one is able to construct a chronological sketch of changes in smithed products and the techniques used to make them.[35]

Metal goods, some in copper, others in iron, were well integrated into early central African society. From the ninth to fifteenth centuries, metalworkers produced wares in two distinct categories—charms and ornaments worn on the person, and implements used in war, food production, and craft work. Early in this period, of the nine types of products worked in metals, three were body ornaments and six were tools or weapons. After the eleventh century, when the graves indicate a marked increase in wealth and stratification, more specialization in metal products was in evidence. There were ten different types of body ornaments rather than three, and thirteen types of implements. Throughout what seems to have been a long period of relatively steady economic growth in this locale, metals continued to be important to society, not only in generating wealth but also as vehicles for storing and displaying it.

That these graves contained both copper and iron goods poses some intriguing questions about the chronological and technological relationships of the two metals and also about blacksmiths' skills in working them. Not surprisingly, it is copper, laden with all sorts of implications about metallurgical development and long distance trade,[36] that has drawn the most intense interest from researchers. Time periods for the Upemba burials are bracketed and characterized according to the relative presence or absence of copper goods, presumably to set the stage for the appearance of the famous cast copper ingot currency used in regional trade by the eleventh century.[37] It is likely, however, that smithed copper and iron goods circulated regionally, possibly as recognized currencies, before the cast copper ingots did. Such earlier currencies would include small units of bar metal and metal wire, some perhaps transformed into body ornaments for storing wealth, and standard iron blades of one kind or another. Whether copperworking preceded ironworking or not and where is still an open question, though technical analyses of the Upemba goods indicate that outside the Copperbelt, techniques developed for working with iron were subsequently applied to copper.[38] What this phenomenon suggests is not necessarily that metalworkers failed to specialize but that diversification was their chosen market strategy, with some blacksmiths adapting their skills to each type of metal.

The iron products in particular present a wide range of environmentally and socially adapted tools—hoes for cultivation, harpoons and fishhooks for fishing, hatchets and knives for clearing land, for construction and woodworking, and spears and knives for hunting and warfare. One would have to turn a blind eye to the forms of these objects to assume that they were all made in a few local smithies. Hoe blades, for example, show standardized forms despite having been found in different locales and throughout the entire period. Moreover, the blacksmiths who made these goods had mastered standard blacksmithing techniques by at least the thirteenth century.[39] The variety of smithing techniques they used suggests regional workshop differences and specialties. Equally important are the anomalies, certain elaborate and unusual luxury products made of iron which suggest a more distant, foreign origin.[40] Interestingly enough, no evidence was found of any smelting, and no evidence of the kind of tools that would have been used to refine iron blooms. The only smithing tools among the grave goods were anvils and a hammer of small dimensions,[41] the kind of tools that would be used in a

Map 1 : Major Central African Archaeological Sites mentioned in text

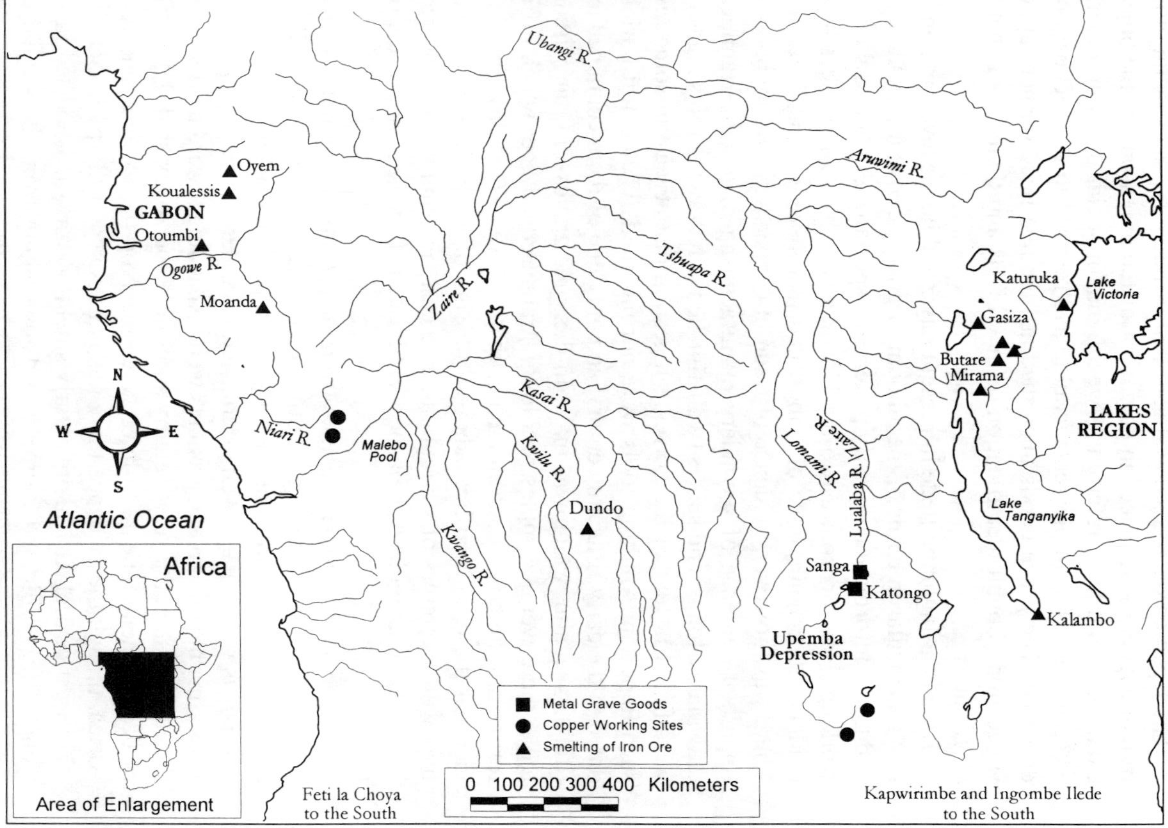

finishing forge, not for refining blooms or turning out hoe or hatchet blades. It is probable, therefore, that iron was traded; local resources of fish and perhaps salt could have been exchanged for iron in early networks that later grew to include copper as well.

The Upemba burials suggest also that different types of metal objects, in iron or copper, were placed in graves for different reasons. Some objects, for example, could have referred to wealth and/or social position; others, such as certain tools or implements, could have referred to the occupation of the deceased. The grave goods unearthed at Ingombe Ilede further support this suggestion, though identification of the objects and their uses is not always clear cut.

Burials at Ingombe Ilede in present day Zambia, far fewer in number and less spectacular in terms of the amount of grave goods they yielded, provide additional information on early patterns of specialization and trade in metal-working. Here, in contrast to the Upemba sites, there was clear habitation evidence, but again, none of it indicated that there was any local smelting. There was, however, evidence of trade in copper wire and ingots, and very important evidence of secondary manufacturing of metals. Hammerheads, drawplates, tongs, and spikes (well-finished wiredrawing tools made of iron) were found in four of the graves, suggesting that at least in some instances such important smithing tools were not inherited. The central burials were dated to the early fifteenth century, and only one of them contained an identifiable skeleton, which was male. In the same graves with the wiredrawing tools, there were examples of a long, straight-edged hoe, which differed in form from the hoes found in the occupation levels. That difference, and the fact that these unusual hoes showed no sign of having been used, suggests they were examples of a circulating hoe currency, and that they were placed in the grave as indicators of the wealth of the deceased, not of his occupation.[42]

A general pattern that emerges from these ironworking remains in central Africa is one of increasing wealth and social complexity, especially around the tenth and eleventh centuries. By that time, ironworkers and other artisan specialists were clearly well established in certain places, regional trade networks that included metal products were expanding toward indirect links with the east coast, and a very wealthy stratum of society existed.[43] A similar line of development appears to have taken place on the western side of central Africa, though the evidence there has not yet been fully published in detail.

In west central Africa, other early smelting workshops existed in areas of Congo-Brazzaville and Zaire, with dates ranging from the second century B.C.E. to the fourth century C.E.[44] Here in the west, too, at sites in and around the Niari Valley, there were areas where both copperworking and ironworking were carried out. At Mindouli, smelting furnaces were usually located on hilltops in clusters of three, for smelting iron and sometimes copper. The oldest one dated so far was from the eleventh century, and two others were from the thirteenth to fourteenth century. Closer to Malebo Pool, the site at Mpassa included a copper furnace near a copper ore deposit, the furnace dated to the thirteenth century.[45] Very little archaeological work has been carried out in Angola. Some traces of iron smelting which date to the tenth century were unearthed near Dundo in the northeast, and iron and potsherd remains from the nearby Tchibaba mine were dated to the eleventh century.[46]

The early history of settlements near and along the rivers of the Zaire basin is still hardly known at all, and so one can only speculate on the history of ironworking there. Survey work has just begun in the past two decades, and has focused on some of the major rivers themselves to establish general patterns and chronologies of settlement. Most of the evidence gathered is in the form of ceramics, and they show that early communities in the equatorial forests were involved in extensive networks of contact and communication during the first half of the first millennium C.E.[47] Precisely when these early communities started to use and then produce iron, however, is still not known.

IRONWORKING AND POLITICAL ORGANIZATION, AFTER CA. 1100—1400

What the archaeological record shows clearly is a broad trend toward increasing social and economic complexity in central Africa, with higher levels of wealth and social stratification occurring in some areas around the beginning of the second millennium. As this general trend continued, the early second millennium became an era of increasing political competition and centralization, though on a limited scale. In writing the histories of some of the earlier central African kingdoms, historians have sometimes asserted that ironworking played a major role in the formation of centralized government. At first glance, this might seem like a reasonable assertion to make. After all, ironworkers and iron imagery crop up in some oral traditions that tell of the founding of polities or ruling dynasties, and ritualized ironworking gestures were sometimes featured in the investiture ceremonies of leaders.[48] But not only are these images and rituals very different in the kinds of references they make to ironworking, they are quite possibly anachronistic and not to be taken at face value. Rarely, however, are they examined critically, and with chronological issues in mind.[49]

Also unexamined is evidence of ironworking itself. Instead, faulty generalizations and assumptions about it have sometimes been accepted in explanations of central African state formation. But just as a closer and more critical examination of the linguistic and archaeological evidence showed that the roles of ironworkers in early Bantu communities were much more complex and nuanced than had been assumed, a closer look at what evidence there is for the roles of ironworkers in state formation suggests that no single, easily summarized explanation or model will ever suffice. Moreover, when oral traditions are reviewed systematically alongside other data from archaeology and linguistics, doubt is cast on any conclusions based solely on the former. That there may well be anachronisms in the traditions only underscores the importancc of consulting contemporaneous primary data, and preferably more than one kind.

In writing the political histories for central African states, historians have tended to sidestep these problems rather than directly confront them. While acknowledging that ironworking must have been important, their studies contain only scattered allusions to ironworking and ironworkers, and in the most general of terms. When an explanation is attempted, it is usually based on some sort of ecological determinism. Control over and exploitation of iron ore resources have been proffered as factors influencing the early phases of political centralization, though rarely is there further explanation or evidence as to how or why this would have occurred. The most systematic exploration of the relation between smelting and state formation resulted in a counterexample.[50] Thus far, then, there is little or very unreliable evidence for supposing that iron ore resources were a major factor in state formation.

Nevertheless, some grand claims that were once made continue to be widely disseminated. In one outdated but still prominent example, scholars described the lands north and south of the lower Zaire as the lands of "the blacksmith-kings." Referring specifically to the founding of the Kongo, Loango, and Ngoyo kingdoms, this characterization was based on thin evidence. At its core was the presence of smelting at sites like Mindouli, but the rest was speculation. Metalworking in the upper Niari River area supposedly offered superior weaponry and therefore a military advantage to migrating warriors, the putative founders of these kingdoms before the fifteenth century.[51] Just what those weapons were or who those warriors were seems not to have mattered. In this case, an oft-repeated migration and military conquest model was made to serve as an explanation for seemingly portentous references to smiths in oral traditions of these kingdoms.

The Luba kingdoms are another well-known example where iron ore deposits and smelting supposedly were major factors in state formation. Reefe's historical study appears to have been generally accepted, notwithstanding his failure in making his case about iron. His thesis was that the heartland of the so-called Luba empire was rich in salt and iron ore, and that those resources accounted for why the area became a major emporium supplying extended regional trade routes in the seventeenth and eighteenth centuries. However, his evidence regarding iron production was thin and anachronistic. For this period, which he characterized as "empire emergent," he relied on nineteenth-century evidence for both iron production and trade.[52] The oral traditions that associate smelting with Kalala Ilunga, dated by Reefe to approximately the seventeenth and eighteenth centuries, have not been corroborated by any other data. Furnaces that operated in the Luba heartland have not been systematically surveyed or dated to find out how long ago they may have been in operation. Moreover, Reefe did not even bother to consult published geological reports that would have tempered his assertion that iron ore deposits were relatively rare in this part of the Zaire basin. In other words, Reefe did not pursue contemporaneous evidence for ironworking during this two-century period, and therefore he did not prove his thesis.

Miller made a more modest and better-supported claim about the importance of iron to the early Mbundu kingdoms in Angola. He pointed out among other things that there were numerous sources of iron ore in the area. Noteworthy, though, were the richest ore deposits, especially those in the Nzongeji River valley, where at least two kingdoms had been created by the fifteenth and sixteenth centuries. Relying on several sets of oral traditions, some recorded in the seventeenth century, Miller recounts that the legendary founding king of the Samba reputedly made it possible for the Mbundu to make axes, hatchets, knives, and arrowheads for the first time. But he wisely refrains from interpreting this episode literally.[53] It might refer not to the introduction of ironworking but to some type of improvement or other change in smithing associated in the oral traditions with this time period. Most significantly, it stresses that it was blacksmiths' skilled work that was important, and not so much the control over iron ore.

Nevertheless, Miller cautioned that kingdoms in this area were probably not founded by smiths. Instead, he proposed that it was mainly the playing out of political maneuverings and rivalries among individuals and groups that must account for the rise of kingdoms. In particular, it was new forms of social and political institutions based on relationships other than kinship that were crucial in facilitating the organization of them.[54] Various oral traditions and testimonies, along with written accounts and linguistic data, leave little doubt that certain groups of ironworkers were well known and that certain iron-rich districts were very productive at certain times in the history of the Mbundu kingdoms. But this is to be expected in societies run by men who desired or depended on so many types of iron products. In other words, it makes sense that newly forming kingdoms would include prominent ironworkers and perhaps some resource areas, but it does not automatically follow that iron would explain the rise of the kingdoms themselves.

Ironworkers were important figures in precolonial central African politics, but what remains to be established is precisely in what ways and why, case by case. So far, much of the evidence historians have cited for ironworking after 1400 is in the form of clichés from oral traditions. Corroboration by hard evidence is not yet possible since there has been hardly any archaeological work carried out on iron production in any of the major precolonial central African states. The most important exception to this is the

Tio kingdom, where archaeologists have examined and analyzed the remains of intensive smelting operations on the Mbé plateau in the sixteenth century. Even more striking, though, is the absence of any mention of ironworking in the Tio kingdom's oral traditions.[55] In this case, the contradiction proves to be much more valuable to historians than any real or imagined similarities among the ironworking clichés of oral traditions. It helps to identify the problem, and it sounds a note of caution. Where there are such clichés, hard evidence to support them is so far lacking; where there is substantial hard evidence for iron production, ironworking clichés are not present. Hence there is a great need for collaborative projects involving archaeologists and historians in the study of ironworking and the rise of central African kingdoms and societies between ca. 1100 and 1900. Until then, all that can be said is that when ironworkers are finally integrated into the political histories of the Zaire basin region, there will be no single answer to the question of the roles they played and no simple pattern to them.

THE SETTING: WEST CENTRAL AFRICA IN THE NINETEENTH CENTURY

Having established that ironworking in central Africa over the past few centuries is still poorly understood, a regionwide survey and analysis of it during the nineteenth century can go a long way toward filling an important gap in precolonial history. This book focuses on ironworkers in west central Africa, a region defined here by the entire Zaire River basin above Malebo Pool, including the basins of the Kasai, Kwilu, and Kwango Rivers, with occasional forays into adjacent areas such as the Ogowé basin in Gabon. The peoples are generally speakers of Bantu languages in the B, H, C, D, L, M, and K groups,[56] though there are some exceptions. Because peoples and their work were neither geographically fixed nor socially isolated, the spatial and linguistic boundaries of this study are necessarily imprecise. The emphasis is on the region inland from the western coastal areas, that is, the parts of west central Africa that stayed relatively well insulated from direct international trade contacts until the nineteenth century.

The craft tradition ironworkers created in central Africa, as evidenced by the archaeological record, shows clearly how iron and iron products were integrated into societies there well before the 1800s. Having a more substantial body of contemporary material evidence from museum and private collections, the additional lexical data, and the occasional reliable written account, along with other evidence generated in the nineteenth century, makes the study of the work during this period a particularly fruitful and rewarding prospect. It offers the opportunity to survey in greater detail the contours of iron consumption and production, and thus also the social position and value of ironworkers in their communities. This period also represents a critical time in the history of the region and the work, for it comes just before major structural and cultural changes were initiated by a variety of colonial policies implemented after the 1920s (see Epilogue).

The nineteenth century, defined here as the period from about the 1820s to the 1920s, was a time when central Africa became more directly drawn into global economic networks. Before then, trade links between the inner Zaire basin and the west and east coasts had been indirect. These new, direct links to international capital came later to central Africa than to other parts of the continent, their impacts thus more concentrated and sharply felt. Moreover, largely because of the nature of Africa's foreign trade, these new links ushered in an era of contradiction, of expanding economic horizons accompanied by social and political upheaval. Ambitious slave and ivory traders, financed mainly by European, Indian, and Omani capital, competed for advantage along the rivers or overland routes between the center of the continent and their bases on the west or east coasts, or north along the Nile. As a result, raiding for slaves and hunting for ivory increased in Africa's interior regions especially after the 1850s, extending the frontiers of international trade further and further inland. Along the axes of commerce, groups of merchant raiders and hunters equipped with firearms converged on central Africa from all sides.

In east central and north central Africa, Muslim economic interests dominated international trade until the 1890s. On the eastern side, the most important of these were Swahili merchants based in Zanzibar, who after mid-century settled in Nyangwe and Kasongo on the Lualaba river just west of Lake Tanganyika. Overland caravan routes connected these fortified towns with the coast, where the ivory and those slaves who survived the journey were sold for export. Ambitious and ruthless merchant-warlords carved out territorial claims in the southeastern portion of the Zaire basin region in order to insure that their caravans would be regularly supplied. The most famous of these was Tippu Tip, who created a state astride the Lualaba and Lomami Rivers after the mid-1870s, using it as a secure base from which to launch raids into Luba and Songye chiefdoms. Another prominent example was the Nyamwezi merchant Msiri who established himself as overlord in the Copperbelt. Local clients and allies protected themselves by serving as these merchants' supply agents.[57]

Other groups of Muslim traders from the Nile region around Khartoum descended on the savannas just north of the Zaire basin, also to purchase ivory and slaves. By the middle of the nineteenth century this trade had undergone a fundamental change. Merchants based in Khartoum formed slave-trading companies, which operated on a much larger scale than ever before, raiding for slaves rather than simply purchasing them. In this case too, territories were carved out and divided up among the companies around the stations they established for their slave-raiding and bulking operations. Company zones extended into what is now northeastern Zaire, and although over time the merchant groups changed in how they were composed and capitalized, the savanna and northern forest area continued to be an important source of slave and ivory supply until the 1890s. By that time, Muslim commercial communities, some originating in Egypt and Sudan, others in East Africa, had managed to forge alliances with leaders of the Azande principalities and establish residency there.[58]

On the Atlantic side, formal abolition of the transatlantic slave trade cast the repercussions of increased foreign commerce rather differently. A transition from exporting slaves to the products of "legitimate trade" prompted shifts in the pattern of supply networks between the coast and the interior. In Angola, many of the Luso-African agents of the slave trade eventually withdrew back to the coast while the opportunities for supplying other commodities to the world market were taken up and exploited by a younger generation of local central African entrepreneurs. Using European credit, they established new companies on the coast and inland. Building on an earlier model for commercial organization based on patron-client relations, merchants and their clients took advantage of declining slave prices to build unfree labor forces for porterage and production. The most lucrative commodities of the revamped export sector were wax and ivory in the 1840s and 1850s, and rubber after 1870, all of them natural products. Wax and rubber, though potentially renewable, were not cultivated on a large scale for the long term, and so all of these commodities tended to be, like slaves and ivory, frontier products. Caravans organized by Ovimbundu merchants operated between the plateaus inland from Benguela and the Lunda-administered territories in the upper Kasai basin which were rich in wax and ivory. After the 1840s, Cokwe entrepreneurs became the most prominent and well-armed producers of export commodities in the interior, supplying caravans headed westward toward the port of Luanda. Cokwe military and economic power was such that by the late 1870s they were expanding into western Lunda, and seized its capital in 1886.[59]

The Zaire River trade was also transformed in the nineteenth century by changes in the Atlantic system. Slave exports from ports near the mouth of the Zaire River continued until the 1860s; ivory exports grew with the rise in ivory prices on the coast after 1830. Along the middle Zaire, merchant groups extended their market reach deep into the river systems. They were able to build and maintain trading zones by establishing strategically located settlements either near the mouth of a tributary or at a natural barrier to through river traffic. Some of these settlements grew into large towns by the 1850s. Competition between firms was fierce, not only for the supply of ivory for export but also for slave labor, which was owned and exploited on a greater scale than ever before. The most promi-

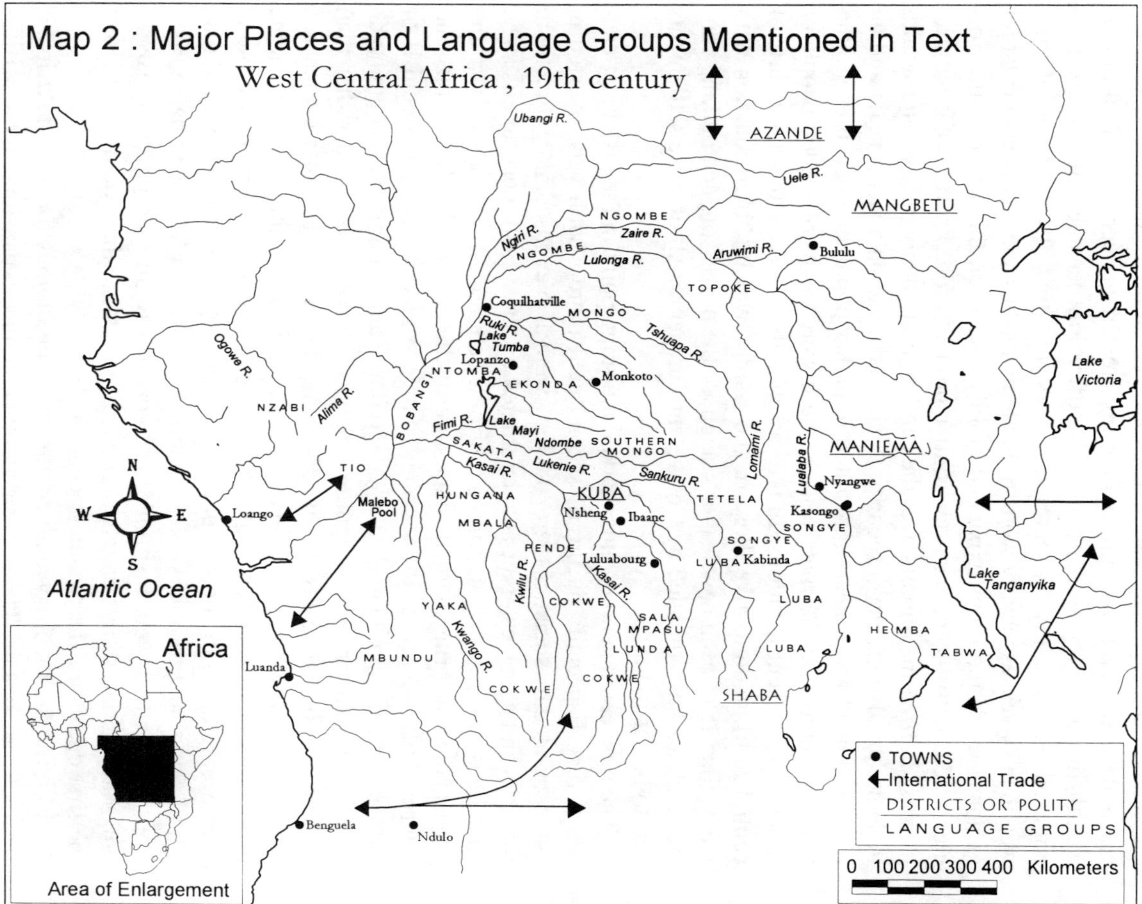

Map 2 : Major Places and Language Groups Mentioned in Text

West Central Africa , 19th century

nent merchant group was the Bobangi, who operated between the mouths of the Kwa and Ubangi Rivers.[60]

However different the origins of these merchant groups and the markets and supply areas within which they operated, the major immediate effects they had on central African societies were very similar. An increase in external trade and merchants' profit margins created greater concentrations of wealth in the region's economies. Stratification became much more pronounced with wealthy leaders and traders investing at least some of their wealth in slaves, especially women. Settlements thus tended to become larger as houses and chiefdoms expanded rapidly. However, the nature of this foreign trade, with its reliance on frontier products for export, along with an increase in firearms imports over the century, brought not only greater wealth to some but also greater insecurity to all. Slave use on a larger scale and an increased need for protection resulted in higher population densities around trading centers. On the river system, these were towns of 10,000 people or more at permanent locations. In the savanna, these were either the courts of centralized kingdoms like those of the Lunda, which shifted location with the beginning of each new leader's reign, or the processing and bulking centers such as those founded by Cokwe entrepreneurs. All of these settlements had to be fortified.[61]

The nineteenth century thus marked a turning point in the history of central Africa. Especially during its second half, more wealth was invested in and generated by societies there, but they were also exposed more directly to the risks and damages of a trade dominated by slaves and ivory and its steep social costs. The rising importance of other commodities, such as wax and rubber, did not bring with it much improvement over the disruptions and intermittent violence of the slaving frontier. And when King Leopold II of Belgium superimposed his brutal Congo Free State administration onto much of the region after 1885, ongoing upheavals were exacerbated rather than diminished. In the 1890s, Leopold's army, the Force Publique, launched a series of military campaigns against Nyamwezi and Swahili overlords in the east, claiming their lands by right of conquest. But revolts and mutinies among the lower ranks of the Force Publique soldiers made genuine and widespread pacification next to impossible. Moreover, the forced labor and harsh taxation policies that were instituted by Leopold's regime provoked uprisings, emigrations, and spontaneous eruptions of resistance that continued through the first decade of the twentieth century.[62]

Even though the nineteenth century was tumultuous for central Africa, it was not cataclysmic. There was general economic expansion, an intensification of change, and episodic insecurity, but society was not reordered, nor were its basic cultural values fundamentally challenged or undermined. Ironworkers were able to maintain their craft tradition throughout, for the rising levels of wealth meant that demand for iron products flourished and grew.

Indeed, that is what this book demonstrates. In the chapters that follow, I show, among other things, that despite several centuries of iron imports along the west coast, ironworkers inland continued to smelt iron ore and that iron was highly valued as a metal. I further show that the greater volume of trade brought about increased iron currency circulation and that the expansion of ivory hunting, slaving, and armed violence stimulated innovation and production of personal weapons. Finally, I show that the rise and fall of rich warlords and big men resulted in conspicuous consumption of the luxury iron goods that glorified their powerful positions. Thoughout this time, ironworkers kept on working, though they did not all experience the same sorts of changes, nor did all of them respond to these new social and economic conditions in the same way. Some left the occupation voluntarily, others were forced out of business. Many others found ways of navigating through the upheavals, enriching themselves by taking advantage of new opportunities and growing markets. In short, iron and iron goods remained culturally integrated into central African society, and so it may very well be that the nineteenth century was the apex of this craft tradition, and the greatest peak production period ever for ironworkers regionwide.

NOTES

[1] J. Forbes Munro, *Africa and the International Economy 1800–1960* (London: Dent and Sons, Ltd., 1976), p. 28. It is not clear whether he was referring to mineral ores or to native metals or smelted metals; the use of the terms "mineral" and "metal" is not a mere semantic detail, but instead carries profound economic implications regarding trade and production.

[2] Laterite is the preferred general term, though the terms limonite and brown hematite are sometimes used. Laterite contains limonite (hydrated iron oxides) as well as other minerals such as clays and quartz. I would like to thank David Killick for straightening me out on this point.

[3] T. Masui, *Guide de la Section de l'Etat Indépendant du Congo à l'Exposition de Bruxelles-Tervuren en 1897* (Brussels: Veuve Monnom, 1897), p. 277.

[4] This point was brought home to me in the two experimental iron smelts I took part in at the University of Toronto, directed by J.E. Rehder, Senior Research Associate, Department of Metallurgy and Materials Science. The first one, on 10 May 1988, was successful, resulting in portions of a bloom with 1.1–1.2 percent carbon content. The second experiment on 18 May, under presumably the same conditions, was not successful.

[5] For a more complete discussion of these issues, see David Killick, "The Relevance of Recent Iron-Smelting Practice to Reconstructions of Prehistoric Smelting Technology," *MASCA Research Papers in Science and Technology* 8, 1 (1991).

[6] Lucien Cahen, *Géologie du Congo Belge* (Liège: Vaillant-Carmanne, 1954), pp. 538–40.

[7] Cornet, and other more recent French speakers, sometimes used the term "oligiste," which is a synonym of the more common term hematite. I am grateful to Dr. Anthony Williams-Jones, Chair of the Department of Earth and Planetary Sciences, McGill Uni-

versity, for sorting out this iron ore terminology for me. Magnetite and hematite are rich ores often found in itabirite deposits, which were of great interest to European geologists. Itabirites (banded ironstones) appear conspicuously in the landscape as mountains with exposed layers of rich iron ore and quartz, and they were the primary focus for colonial geological surveys of significant iron deposits. They were found concentrated especially in the northeast between the upper Uele and Aruwimi Rivers, in the northwest near the bend in the Ubangi River, in an area between the upper Kasai and Sankuru Rivers, around the lower Zaire, and also just north of there near the upper Ogowé River. Cahen, *Géologie du Congo Belge*; Idem, *Esquisse tectonique du Congo Belge et du Ruanda-Urundi* (Liège, 1952); Idem and J. Lepersonne, *Carte Géologique du Congo Belge et du Ruanda-Urundi* (Liège, 1951); and B. Guillot, "Note sur les anciennes mines de fer du pays Nzabi dans la région de Mayoko," *Cahiers ORSTOM* 6, 2 (1969).

[8] Jules Cornet, "Les Gisements Métallifères du Katanga," *Bulletin de la Société Belge de Géologie* (Bruxelles) 17 (1903), p. 41.

[9] Of the smelting operations in the Zaire basin where it is known for certain what type of ore was used, at least three were from itabirite deposits, one of them hematite and at least one magnetite. See D. Schwartz, R. Deschamps, and M. Fournier, "Un site de fonte du fer récent (300 bp) et original dans le Mayombe congolais: Ganda-Kimpese," *Nsi* 8/9 (1991); Francis Van Noten and E. Van Noten, "Het Ijzersmelten bij de Madi," *Africa-Tervuren* 20, 3–4 (1974), and Guillot, "Note." The field demonstration referred to here was commissioned and observed by the Van Notens in northeastern Zaire. In this case, the itabirite formation was made up of magnetite ore banded with quartz. The smelt was successful without the need of adding a flux since the bands of quartz supplied the necessary gangue for iron reduction in the bloomery process.

[10] The iron content could range between 27 and 52 percent; silicon content varied between 21 and 54 percent. See David J. Killick, *Technology*; and Ingo Keesman, Johannes Preuss, and Johannes Endres, "Eisengewinnung aus laterischen Erzen, Ruki-Region, Provinz Equator/Zaïre," *Offa* 40 (1983), pp. 183–90. Among many examples, we encountered such ore identification in Lopanzo, Equateur; Cornet mentioned it for areas in Shaba; and Killick noted it in Malawi. Kriger, field notes, Lopanzo, 1989; Cornet, "Gisements," p. 41; Killick, *Technology*, pp. 115–17.

[11] That is, assuming that the ores collected were indeed the ones smelted. The ores were not uncovered in precisely the same place as the slags, but were collected in the vicinity and were assumed to have been the ones that produced the slag heaps. High aluminum oxide content of both the slags and ores suggest the two were linked. Keesman, et al., "Eisengewinnung." Killick recorded a similar finding in Malawi. Killick, *Technology*, pp. 215–16.

[12] R.F. Tylecote, "Furnaces, Crucibles, and Slags" in *The Coming of the Age of Iron*, eds. T. Wertime and J. Muhly (New Haven: Yale University Press, 1980), pp. 211, 216–18.

[13] Kriger, field notes, Lopanzo, Zaire, July–August 1989. See also Georges Célis, "Fondeurs et forgerons Ekonda (Equateur, Zaire)," *Anthropos* 82 (1987).

[14] F. and E. Van Noten, "Het Ijzersmelten"; F. Van Noten, "Ancient and Modern Iron Smelting in Central Africa: Zaire, Rwanda, and Burundi," in *African Iron Working—Ancient and Traditional*, eds. R. Haaland and P. Shinnie (Oslo: Norwegian Universities Press, 1985); and Manfred Eggert, "The Current State of Archaeological Research in the Equatorial Rainforest of Zaire," *Nyame Akuma* No. 24/25 (December 1984). The Mbandaka furnace dated to the late seventeenth to eighteenth centuries.

[15] For the most recent general discussion of this issue, see Miller and Van der Merwe, "Early Metalworking," pp. 19–28.

[16] See e.g. J.E. Rehder, "Primitive Furnaces and the Development of Metallurgy," *Journal of the Historical Metallurgy Society* 20, 2 (1986).

[17] See Bruce Trigger, *A History of Archaeological Thought* (Cambridge: Cambridge University Press, 1989), pp. 252–54.

[18] Temperatures in a bowl furnace used near the middle Zaire reached the range of 1200–1300°C. See C. Kriger, log of iron smelt, Zaire, 2 August 1989, in Idem, "Ironworking in 19th century Central Africa" (Ph.D. diss., York University, Canada, 1992) and published in Kyle Ackerman, David Killick, Eugenia Herbert, and Colleen Kriger, "A Study of Iron Smelting at Lopanzo, Equateur Province, Zaire," *Journal of Archaeological Science*, forthcoming 1998. For experimental iron smelts with small shaft furnaces, where temperatures of 1600°C were reached, see R.F. Tylecote, J.N. Austin, and A.E. Wraith, "The Mechanism of the Bloomery Process in Shaft Furnaces," *Journal of the Iron and Steel Institute* 209 (May 1971).

[19] Again, I would like to thank David Killick for his help in clarifying these technological problems.

[20] Rehder, "Primitive Furnaces."

[21] P. de Maret and F. Nsuka, "History of Bantu Metallurgy." See also the critiques of Bantu expansion theories listed in footnote 12 of Chapter 1.

[22] See Miller and Van Der Merwe, "Early Metalworking," p. 10; and B. Clist and R. Lanfranchi, eds. *Aux Origines de l'Afrique Centrale* (Libreville: CICIBA, 1991), pp. 181–83.

[23] Recent work on this question, using more complete data than Guthrie, shows that root forms once thought to have indicated proto-Bantu knowledge of iron most likely meant "stone" rather than the metal "iron": **-tádè* and **-bùè* are shown convincingly to have probably meant "stone"; **-yúmà* probably meant "thing"; **-gèdà* shows no clear, convincing evidence that its original meaning was "iron." P. de Maret and F. Nsuka, pp. 44–49.

[24] For an example of how promising such research can be when data from linguistic analyses are used with other primary data, see Mary Allen McMaster, "Patterns of Interaction: A Comparative Ethnolinguistic Perspective on the Uele Region of Zaire ca. 500 B.C. to 1900 A.D." (Ph.D. diss., U.C.L.A., 1988).

[25] Radiocarbon dates, usually taken from samples of charcoal, are not as straightforward as might at first appear. In particular, one thorny problem has had to do with the charcoal. Not only is it important to note the context and associated materials of charcoal samples, but the charcoal itself may be problematic if it was made from old wood. If so, the radiocarbon date would be correct for the age of the wood but not for the age of the furnace or slag. For more on this issue, see the very useful discussion in Bernard Clist, "A Critical Reappraisal of the Chronological Framework of the Early Urewe Iron Age Industry," *Muntu* 6 (1989).

[26] See my survey of the archaeological reports in C. Kriger, *Ironworking*, pp. 102–114, 120–21, 361–62.

[27] These "before present" dates are uncalibrated.

[28] M.-C. Van Grunderbeek, E. Roche, and H. Doutrelepont, "L'age du Fer Ancien au Rwanda et au Burundi. Archéologie et Environnement," *Journal des Africanistes* 52, 1–2 (1982), pp. 19–24; B. Clist, "Early Bantu Settlements in West-Central Africa: A Review of Recent Research," *Current Anthropology* 28, 3 (June 1987), p. 381; and Clist and Lanfranchi, *Aux Origines*, p. 207.

[29] L. Digombe, P. Schmidt, V. Mouleingui-Boukosso, J.-B. Mombo, and M. Locko, "Gabon: The Earliest Iron Age of West Central Africa," *Nyame Akuma* 28 (April 1987).

[30] For a masterful summary of the evidence to date for all of sub-Saharan Africa, see Miller and Van der Merwe, "Early Metalworking." The Gabon evidence has been inte-

grated with other new data into a convincing overview for the early development of iron metallurgy in west-central and west-southern Africa in James Denbow, "Congo to Kalahari: Data and Hypotheses about the Political Economy of the Western Stream of the Early Iron Age," *The African Archaeological Review* 8 (1990).

[31] J.D. Clark, *Kalambo Falls Prehistoric Site*, 2 vols. (Cambridge: Cambridge University Press, 1969 and 1974), Vol. I, p. 194.

[32] At the Kapwirimbwe site east of Lusaka, furnaces were dated to between 400 and 500 C.E. At another site near Lusaka, intensive smelting took place from 800 to 1100. Sometime after the 1300s, perhaps no earlier than the nineteenth century, smelters operated tall shaft furnaces in parts of northern and eastern Zambia, many of which are still extant. D.W. Phillipson, *The Later Prehistory of Eastern and Southern Africa* (London: Heinemann, 1977), pp. 134, 173, 176. See also J.H. Chaplin, "Notes on Traditional Smelting in Northern Rhodesia," *South African Archaeological Bulletin* 16 (1961).

[33] Iron and copper smelting evidence at the site of Naviundu indicates they were done at the same time, ca. 1500 bp. Pierre de Maret, "Recent Archaeological Research and Dates from Central Africa," *JAH* 26 (1985), p. 137.

[34] Another important copper production center was located at Kipushi, where copper smelting and casting of copper ingots was dated to between the ninth and fourteenth centuries, with the size of ingot very large by the early 1300s. See F. Van Noten, *The Archaeology of Central Africa* (Graz: Akademische Druck- und Verlagsanstalt, 1982), pp. 91–92; Michael Bisson, "Copper Currency in Central Africa: The Archaeological Evidence," *World Archaeology* 6, 3 (1975); Idem, "The Prehistoric Coppermines of Zambia" (Ph.D. diss., University of California at Santa Barbara, 1976).

[35] Jacques Nenquin, *Excavations at Sanga, 1957* (Tervuren: MRAC, 1963); Jean Hiernaux, Emma de Longrée, and Josse de Buyst, *Fouilles Archéologiques dans la Vallée du Haut-Lualaba. I. Sanga, 1958* (Tervuren: MRAC, 1971); Pierre de Maret, *Fouilles Archéologiques dans la Vallée du Haut-Lualaba. II. Sanga et Katongo, 1974* Parts I and II. (Tervuren: MRAC, 1985). For a survey of the smithing techniques, see Kriger, *Ironworking*, pp. 157–58, 394–95.

[36] The rich reserves of copper in the Copperbelt of central Africa begin about 250 km south of the Upemba depression.

[37] See J.A. Schoonheyt, "Les Croisettes du Katanga," *Revue Belge de Numismatique* 137 (1991); Pierre de Maret, "Histoires de Croisettes," in *Objets-signes d'Afrique*, ed. Luc de Heusch, (Tervuren: MRAC, 1995); Idem, "L'Evolution Monétaire du Shaba Central entre le 7 et le 18 siècle," *African Economic History* 10 (1981); and Bisson, "Copper Currency."

[38] S. Terry Childs, "Transformations: Iron and Copper Production in Central Africa," *MASCA Research Papers in Science and Archaeology* 8, 1 (1991). See also S. Terry Childs, William Dewey, Muya wa Bitanko Kamwanga, and Pierre de Maret, "Iron and Stone Age Research in Shaba Province, Zaire: An Interdisciplinary and International Effort," *Nyame Akuma* 32 (Dec. 1989).

[39] These include the formation of standardized midribs for different types of blades, and the techniques of heat or pressure welding, splitting, punching, cutting, engraving and/or stamping, wiredrawing, twisting, and plaiting. Kriger, *Ironworking*, pp. 394–95. See also the metallographic analyses of some of the Sanga burial objects, where it was noted that the smiths showed mastery of welding, and that they used pile welding and/or work hardening to strengthen the metal. S. Terry Childs, "Iron as Utility or Expression: Reforging Function in Africa," *MASCA Research Papers in Science and Archaeology* 8, 2 (1991), pp. 59–61.

[40] See Kriger, *Ironworking*, pp. 146–52. Brian Fagan drew the same logical inference with regard to the Ingombe Ilede burials, suggesting that the luxury objects were imported, while utilitarian goods were from local sources. B. Fagan, "Excavations at Ingombe Ilede 1960–2" in *Iron Age Cultures in Zambia*, B. Fagan, D.W. Phillipson, and S.G.H. Daniels. 2 Vols. (London: Chatto and Windus, 1967 and 1969), Vol. 2, p. 102.

[41] Hiernaux, et al., *Fouilles;* de Maret, *Fouilles*. See also the published data in Graham Connah, *African Civilizations* (Cambridge: Cambridge University Press, 1987).

[42] Fagan, "Excavations," pp. 55–161; D.W. Phillipson and Brian Fagan, "The Date of the Ingombe Ilede Burials," *JAH* 10, 2 (1969), p. 202.

[43] De Maret, *Fouilles*, Pt. II, pp. 265–67.

[44] In Mayombe, slag and furnace tuyères dated to the second century B.C.E., and smelting sites in the lower Zaire dated from the first to second century C.E. Other furnaces near Malebo Pool dated to the fourth century. Clist and Lanfranchi, *Aux Origines*, pp. 210, 216.

[45] Clist and Lanfranchi, p. 209. The Mpassa date, 1290 ad +/- 80, is uncalibrated. De Maret, "Recent Archaeological Research," p. 141.

[46] To the west and south of these sites lies the Benguela Plateau, and there the midden at Feti la Choya yielded slag, tuyères, beads, and iron fragments. The occupation period spanned the eighth to the thirteeenth centuries. Van Noten, *Archaeology*, pp. 77–96; Clist and Lanfranchi, pp. 220–21. Quite far south of the Zaire basin, in northwest Botswana, archaeologists have recently uncovered an array of iron and copper objects that can be compared with finds elsewhere, such as those at Upemba. The social and economic complexities represented in the objects indicate that the site of Feti la Choya in Angola was not an anomaly, and that there were societies in the west, too, that had become much wealthier toward the end of the first millennium. See Denbow, "Congo to Kalahari," pp. 141, 161–66.

[47] Manfred Eggert, "Remarks on Exploring Archaeologically Unknown Rain Forest Territory: The Case of Central Africa," *Beiträge zur allgemeinen und vergleichenden Archäologie* 5 (1983), pp. 318–20; Idem, "Imbongo and Batalimo: Ceramic Evidence for early settlement of the Equatorial Rainforest," *African Archaeological Review* 5 (1987), pp. 132–33; and Clist and Lanfranchi, pp. 214–15.

[48] See examples discussed by Herbert, *Iron, Gender, and Power*, pp. 131–47.

[49] An exception is the study by Zdenka Volavka, *Crown and Ritual: Royal Insignia of Ngoyo*.

[50] A study of one West African iron-smelting center showed that although large-scale iron production increased population density, trade, craft specialization, and communitiy stability, it did not lead to political centralization. See Philip de Barros, "Societal Repercussions of the Rise of Large-Scale Traditional Iron Production: A West African Example," *The African Archaeological Review* 6 (1988).

[51] Chapter 11, "The Land of the Blacksmith Kings" in R. Oliver and A. Atmore, *The African Middle Ages 1400–1800* (Cambridge: Cambridge University Press, 1989).

[52] Reefe, *The Rainbow and the Kings*, pp. 7, 84–85, 98.

[53] Miller, *Kings and Kinsmen*, pp. 36, 66–67, 74.

[54] Ibid, pp. 74–76, 86–88.

[55] See Marie-Claude Dupré, "Pour une histoire des productions: la métallurgie du fer chez les Téké Ngungulu, Tio, Tsaayi (République du Congo)," *Cahiers ORSTOM* 18 (1981–2); Idem and B. Pinçon, *Métallurgie et Politique*.

[56] Guthrie classifications, amended by the Linguistics Section at Tervuren. Y. Bastin, *Bibliographie Bantoue Selective* (Tervuren: MRAC, 1975).

[57] Jan Vansina, "Peoples of the Forest," in *History of Central Africa*, 2 Vols., eds. Phyllis Martin and David Birmingham (London: Longman, 1983), Vol. I, pp. 115–16; Thomas Reefe, "The Societies of the Eastern Savanna," in Martin and Birmingham, *History of Central Africa,* Vol. I, pp. 196–99.

[58] Dennis Cordell, "The Savanna Belt of North-Central Africa," in Martin and Birmingham, *History of Central Africa*, Vol. I, pp. 60–73.

[59] Joseph Miller, "The Paradoxes of Impoverishment in the Atlantic Zone," *History of Central Africa*, Vol. I, pp. 151–55; Jean-Luc Vellut, "Notes sur le Lunda et la Frontière Luso-Africaine (1700–1900)," *Etudes d'Histoire Africaine* 3 (1972).

[60] Vansina, "Peoples of the Forest," pp. 114–15; Robert Harms, *River of Wealth, River of Sorrow* (New Haven: Yale University Press, 1981), pp. 24–43.

[61] Vansina, "Peoples of the Forest," p. 101; Idem, *Children of Woot*, pp. 9, 87; Vellut, "Notes sur le Lunda," pp. 72–75; Reefe, "Societies of the Eastern Savanna," pp. 176, 195; Idem, *The Rainbow and the Kings*, pp. 89–90; and Joseph Miller, "Cokwe Trade and Conquest in the Nineteenth Century," in *Precolonial African Trade*, eds. Richard Gray and David Birmingham (London: Oxford University Press, 1970).

[62] P. Ceulemans, *La Question Arabe et le Congo 1883–1892* (Brussels: ARSOM, 1959); Guy de Boeck, *Baoni. Les Revoltes de la Force Publique sous Léopold II, Congo 1895–1908* (Anvers: EPO, 1987); Jean-Luc Vellut, "La Violence Armée dans l'Etat Indépendant du Congo," *Cultures et Développement* 16, 3–4 (1984); Allen Isaacman and Jan Vansina, "African Initiatives and Resistance in Central Africa, 1880–1914" in *UNESCO General History of Africa* (Paris, 1985), Vol. 7; and Jules Marchal, *E.D. Morel contre Léopold II*, 2 Vols. (Paris: l'Harmattan, 1996).

PART II

SOCIAL AND ECONOMIC VALUES OF IRONWORKING: REGIONAL PATTERNS

3

SMELTING IRON: FATHERS OF THE FURNACE

Ambúda zyeédu abuundáá mbógu.
Our forebears used to dig the earth for metal.
Mu daangu dyambúd'éedu mbugu abáabáá Ahunganá.
Long ago it was the Hungana who were the ironworkers.

—Mbala[1]

Ironworkers who specialized in smelting are the subject of this chapter, which I begin with my survey and map of furnace locations for the Zaire basin region during the nineteenth century. These workshop locations are based on written records made mostly by European and North American observers from the 1860s onward, though it can be presumed that each recorded example represents a smelting procedure that was at least a generation old.[2] I then analyze and discuss various technological, environmental, and social factors involved in smelting, and devote separate sections to estimates of furnace outputs, divisions and management of labor in preparation for smelts, and the ways smelters controlled access to skills and hence the numbers of furnaces in operation. The chapter closes with a brief discussion of ritual prohibitions and technology transfer, showing how master smelters exercised monopolies over their skills while also reproducing themselves on a very selective basis.

My survey and analysis of the evidence reveal three paradoxical aspects of ironworking. One is that despite the widespread supplies of iron ore and charcoal fuel available in the region, iron as a metal was scarce and valuable. This important point, that iron was a semiprecious metal, lays the groundwork for Chapters 4 and 5 which focus on heretofore neglected aspects of ironworking—the work of making iron currencies and luxury goods. Another paradox has to do with the complexity of ironworkers' identities. They were part of a multiethnic occupational group with ties to each other based on skill and the potential transfer of skill from one workshop to another. At the same time, they were members of local communities where they harnessed local ideologies to enhance their reputations. The third paradox is that women were denied direct access to iron production and blooms of iron, and were apparently not the major consumers of ironwares, even though their work as the primary agriculturalists throughout the region contributed substantially to demand for iron tools.

EVIDENCE FOR SMELTING WORKSHOPS

Central Africa's well-known foreign trade in slaves, ivory, and various forest and agricultural products was built on an earlier and continuing regional trade, much of it in metal goods. European observers in nineteenth- and early twentieth-century central Africa could not help but notice ironworkers and the many different markets for their products. Some even took the trouble to record descriptions of workshops they saw, noting at times how the work was carried out, and sometimes mentioning which specific markets particular groups of ironworkers had in mind. Useful though these eyewitness accounts are, they are uneven in the terrain they cover and in the level of detail and accuracy they provide about what ironworkers actually did. Most European observers were not very knowledgeable about the processes of smelting and smithing, and their accounts are often very limited and sometimes outright erroneous. They are also limited in time, going back only to the 1860s and 1870s. Still, they usually contain at least some valuable and reliable nuggets of data, such as specific details about the types of equipment being used or aspects of the labor process. And in most cases, these descriptions are for specific, named geographical locations that can be plotted on a map. While such accounts are limited to sites visited by Europeans, and so do not reveal the full geographical distribution of iron workshops (which, especially for smithing, were much more widespread than my maps and survey will indicate), together they at least provide a general picture of where certain ironworkers chose to locate their workshops in the latter part of the nineteenth century.

Other valuable details included in these descriptions vary from references to specific ore deposits, to sizes and features of smelting furnaces, to accounts of the smelting process itself. By piecing these details together with relevant linguistic data such as lexical terms for smelting furnaces, I present in this chapter examples of how the work was carried out in some of these different locales. The picture that emerges does not portray the uniform, static, simple technology assumed by many historians. Instead and in contrast, it reveals the particular forms and features of a technology whose very flexibility spawned a variety of working processes and types of equipment, all of them created and maintained by ironworkers. Taking first the bare essentials of the working process, and then comparing that with what ironworkers did in this or that workshop, reveals very clearly that over time different ironworkers in different parts of the region modified, elaborated, and amended the technology of ironworking and how best to organize the labor for carrying it out. One sees, for example, where and in what ways some ironworkers intensified their production, where and how they tried out certain technological innovations, and how they divided and organized the labor process.

SMELTING WORKSHOP LOCATIONS, CA. 1860–1920

Smelters producing iron in nineteenth-century central Africa were in certain respects relatively unconstrained in choosing where to locate their workshops. In contrast to blast furnace smelting, bloomery smelting does not require heavy inputs of energy, ore, and fuel. Bellows, for example, were operated manually by the smelter's assistants, and as I have already pointed out in Chapter 2, ore of one kind or another was available in much of the region. Smelters, especially those in rainforest and woodland areas, could usually depend on access to sufficient supplies of wood for making charcoal. By identifying and selecting the types of hardwood species that yielded the most effective charcoal fuel,[3] smelters in many areas of the region not only improved the operating efficiency of their furnaces, but they differentiated their needs from those of the household, where other types of wood were used for cooking. All in all, supplies of these essential human and natural resources—energy, ore, and fuel—were plentiful enough throughout the region that smelters were not strictly tied by them to specific geographical sites.

The potential for geographical mobility of central African smelters stands in contrast to an image of ironworking coming from the European context, where blast furnace smelting increased the scale and intensity of iron production and smelters took on new restrictions in how and where the work was done. The blast furnaces that became so common in Europe during the sixteenth and seventeenth centuries were powered by water-driven bellows and had to be permanently located adjacent to flowing water. And while blast furnaces increased the volume of iron produced, they also consumed enormous amounts of ore and charcoal fuel. The problem of deforestation, caused primarily by agricultural expansion but also by reliance on using wood as a general fuel, prompted a search to find a substitute for charcoal in smelting. Coal was a solution developed in the eighteenth century, but as that problem was solved, others arose. Energy demands grew as coal had to be converted to coke, and transportation costs became increasingly important as distances from coal mines as well as from iron ore mines had to be considered.[4] Even this oversimplified chronology shows how the technological paths followed in the history of European iron production were substantially different from African ones.

That central African smelters exercised their relatively greater potential for geographical mobility is reflected in Map 3, which shows the locations of about one hundred smelting workshops during the late nineteenth and early twentieth centuries. In most of these cases, smelters operated some distance away from the primary urban centers of the region.[5] Although obviously it would have made sense to produce their iron very close to major bulking and distribution centers, economic factors were not the only considerations smelters had to face when making decisions about where to work. Social upheaval and personal security could in some cases override the attraction of the most economically convenient locales. The violence of the slave and ivory trade dominated much of the region's main trading arteries in the second half of the nineteenth century. And, although much of that violence was forcibly put down by wars of colonial conquest, it was soon replaced by the violence of colonial forced labor policies and the infamous concessionary system. Thus this period of general economic expansion, while offering more economic opportunity in the way of higher profit margins and greater market demand, created a climate that could be at times inhospitable to independent primary producers of iron. And, depending on what conditions would be at any given place and time, smelters coped with this new climate in different ways.

The threat of raiding and theft was very real and ongoing, but how it might play out was always uncertain. Some ironworkers could rely on the importance of their products to provide sufficient leverage in forming alliances for protection. A chief who ruled over an iron mining district north of the Aruwimi River was described as the only chief in this area who never submitted to the "Arabs" [sic]. His success in resisting conquest by Swahili merchant overlords was attributed at least in part to their reluctance to interfere with the iron production in his domain.[6] But if indeed some slave merchants were so discerning, others might not be, and ironworkers could just as easily find themselves the vulnerable targets of attack. Elderly men in a small village near the Lovoi River in Shaba reminisced to a passing missionary about the days before colonial conquest when they used to smelt iron. Their work came to an end when they were raided by warlords from Kabinda, who carried off all of their tools and threatened them never to take up ironworking again.[7] Fortunate to have escaped with their lives, these particular men took the threat seriously enough to completely abandon their occupation.

Relocating workshops was yet another option, given both the portability of the equipment and the availability of essential resources. Some ironworkers did so, moving to more protected places upriver, inland, or in swampy terrain. It cannot be mere coincidence that in the late eighteenth century, when the middle Zaire basin became a more important supplier in the Atlantic slave trade, several blacksmiths moved 250 km up the Ngiri River away from where it emptied into the Zaire, and founded new settlements. Upriver the ground was low in elevation, crisscrossed by innu-

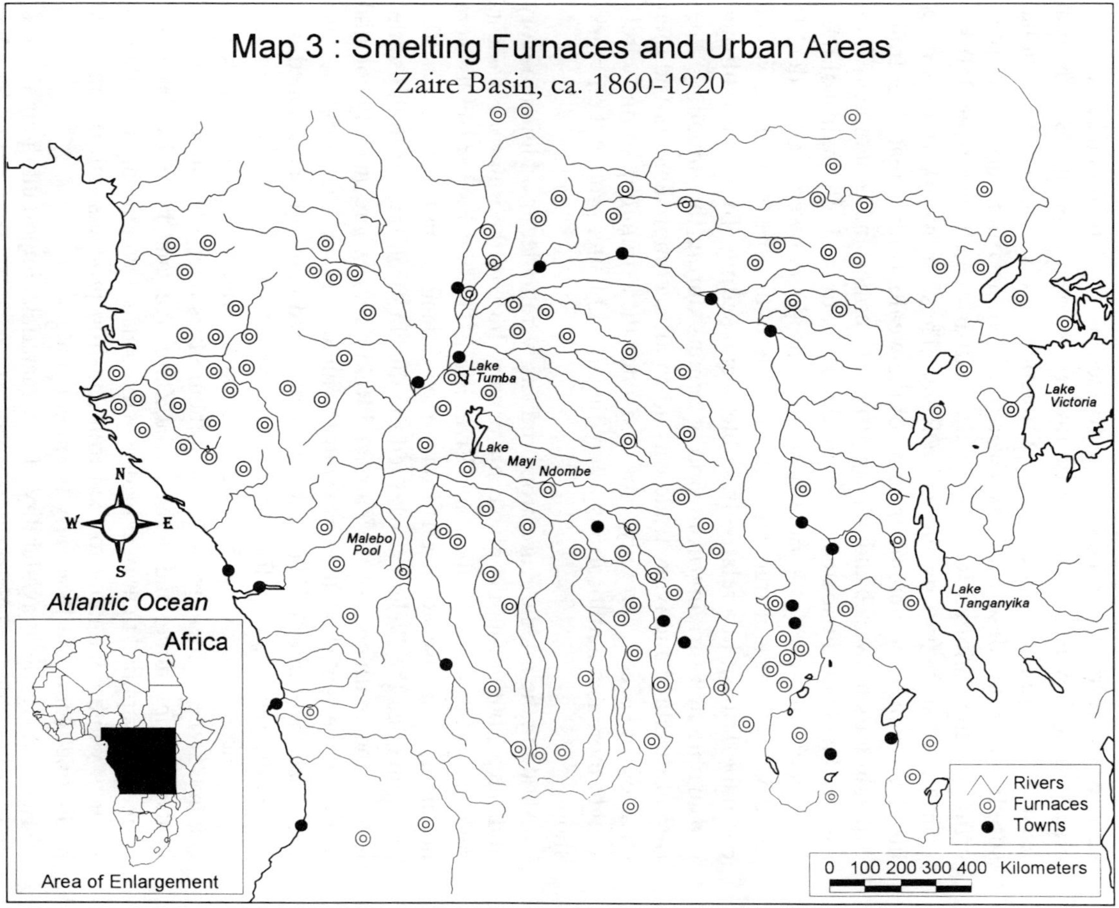
Map 3 : Smelting Furnaces and Urban Areas
Zaire Basin, ca. 1860-1920
Lake Tumba
Lake Mayi Ndombe
Malebo Pool
Lake Victoria
Lake Tanganyika
N
W
E
S
Atlantic Ocean
Africa
Area of Enlargement
Rivers
Furnaces
Towns
0 100 200 300 400 Kilometers

merable streams, ponds, and marshes. In this seemingly remote area they set up shop, contracting with distant smelters to augment their smithing skills with the knowledge of how to produce iron.[8] Their settlements grew into prosperous villages, which during the nineteenth century monopolized smelting around the upper Ngiri and supplied a range of iron products for the territory in the triangle formed by the Ubangi and Zaire rivers. At first they relied on lateritic ores from local deposits. But from at least the mid-nineteenth century, production levels demanded additional supplies of ore; these came by river transit trade from various sources, primarily the Lulonga River near Basankusu, which was over 350 km from the nearest port on the Zaire River. These ironworkers thus drew into their remote communities some of the wealth of the great river trade, expanding its networks and adding growth to hinterland economies. Their success has been attributed, with good reason, to the high quality of their products.[9] But surely the quality of their products owed a great deal to the development and transmission of skills over time, something which can only flourish in a relatively stable social setting. To have established and maintained thriving ironworking centers during the tumultuous period of the slave and ivory trade was no small accomplishment. And their chances for success certainly improved when these founding smiths wisely sought more isolated locations where they could better protect their workshops.

All along the middle Zaire, iron products were traded in addition to ivory and slaves, but during the nineteenth century little of the ironwork was carried out on the shores of the Zaire River itself. The Ngiri River smiths were not an unusual case, for it was ironworkers inland or up tributaries such as the Alima and the Ikelemba who fed the thriving iron trade.[10] In the settlement at the confluence of the Zaire and Ruki rivers, where an early colonial station was established, there was only a single smithy in the 1890s. But one could see there piles of slag, furnace debris, and numerous abandoned iron mines, the remains of a former smelting center.[11] Continual threats of war parties and raids drove smelters away from the area in the early eighteenth century, some of them reestablishing themselves over 120 km to the south, in the swampy hinterland between the Zaire river and Lake Tumba. Once there, they supplied their own leaders with arms for conquering neighboring villages while continuing to feed iron and iron products into the regional trading networks until the early twentieth century.[12]

Hence the general availability of raw materials, important though it was, was not the only or necessarily the primary issue smelters faced when they established their workshops. And once established, smelters could still find themselves forced by new political or social conditions to change where and how they worked. The economic expansion of the nineteenth century, spurred on by external trade, created upheavals and insecurities on an unprecedented scale. But it would be wrong to assume that changes in ironworking during this period were significantly different in kind from change in earlier times. Similar events undoubtedly occurred in the more distant past, though intermittently and localized, and less disruptive overall. The long term continuities in iron production that did take place in central Africa did not unfold at an even and steady pace, but did so during the unpredictable interruptions, lulls, turnings, and lapses of historical time as it was lived. That smelting continued to be practiced throughout the region could not have occurred without ironworkers' strategies for survival of their workshops. And these strategies involved searches for relatively stable or improved working conditions as well as a search for raw materials.

FURNACE CAPACITY AND DEMAND FOR IRON

Smelters not only continued to smelt, but they sought to enhance in a number of ways their abilities to produce iron. The master smelter and his assistants repaired or constructed anew the furnaces they used each dry season, and variations in furnace design reflect historical differences among smelters in their specialized knowledge.[13] The minimum requirement for bloomery smelting is a bowl furnace dug out of the ground with one tuyère and bellows for air supply. This basic type of furnace was common, but so, too, were more elaborate designs. Some examples of furnaces had built-up shafts around them, up to several meters high, and both bowl and shaft furnaces could vary a great deal in their forms and features.

The relationship between the design of a small-scale bloomery furnace and its potential production capacity is neither clear cut nor consistent. It is not simply a matter of how high or deep a furnace is, and successive smelts in the same furnace can produce very different results. In assessing the potential capacity of a furnace, that is, the size of the iron product or bloom it could produce, one must be able to gauge the potential size of the furnace's working area, the area of carbon monoxide atmosphere. In a forced draft furnace with only one tunnel or tuyère bringing in the air supply, the carbon monoxide would be created inside near the mouth of the tuyère. It is possible, then, to increase the area of carbon monoxide, the working area of the furnace, by increasing the number of tuyères and placing them at angles to each other. Furnaces with potentially greater capacity thus may or may not have had shafts or walls built up around them, though it is the latter feature that has most often signaled to researchers that higher volume production was reached. Shafts may have been built up for other reasons, however. For example, a relatively low capacity furnace, that is, a bowl furnace with only one tuyère, might need additional walls built up around it to handle types of ore that were low in iron content and high in gangue, calling for large inputs of ore and fuel as a result. The bloom of iron, though, would have been relatively small. Scale of furnace structures may in some cases have more to do with the particular types of ore that were used rather than the actual volume of the iron product.[14]

Smelters made all sorts of adjustments to furnaces, some of which could have had noticeable effects on how they operated. The locations of furnaces with the potential for producing larger blooms, that is, ones that had shafts and large working areas of more than two tuyères, form a ring around the area of the central Zaire basin. Within the basin itself, and in other places where population densities were lower, the furnaces tended to be those with potentially smaller capacity, bowl furnaces with only one tuyère. This evidence, speculative and incomplete though it is, does suggest a correlation between furnace design and population density, that in more populous areas smelters may have attempted to produce larger blooms for roughly similar labor expenditures.[15] These modifications in furnace design and apparatus might seem insignificant to an economic historian comparing them with the technological advances that took place in preindustrial and industrial Europe. Nevertheless, such innovations would indicate that central African smelters, too, experimented with ways to increase their volume of iron production.[16]

Concrete examples of smelts demonstrate more clearly that furnace design must be viewed as only one of several factors affecting the amount of iron that smelters produced, and the ways these factors interacted could yield surprising results. It is possible to estimate averages of bloomery furnace output based on the data that have been recorded for several specific smelts, and together they underscore the complexity of the process. Output was not simply a function of the exterior height of a furnace, but had more to do with the size of the working area inside. Important, too, was the type of ore. As Table 3-1 shows, bowl furnaces that employed two tuyères and that were used with relatively high grade ore could produce larger quantities of iron in a single smelt than could shaft furnaces using ore of a lower grade. Another factor, this one affecting yields over time, was the working season. Unlike smithing, which could be done year round, mining and smelting were generally confined to the dry periods of the year. High humidity and the absorption of moisture into the charcoal fuel could significantly reduce furnace efficiency.[17] Moreover, moisture in the iron ore itself also increased the burden on the furnace, and bursts of vapor during the smelt could disrupt its workings. Hence even during the dry season smelters took care to reduce any remaining moisture in the ore, especially laterite ore, by breaking it up into smaller chunks and roasting them or drying them out in the sun. Thus the length of the dry season, which varied from two to four months, also had a significant impact on the yearly output of a single smelting team in a particular place.

Table 3–1 Estimates of Iron Output per Furnace per Year: Zaire and Kasai basins, Cameroon, and Malawi

Ore	Furnace	Iron per Bloom	Dry Season	Total per Season*
1. Itabirite	Bowl w/2 tuyères	6 kg	2 mo.	48 kg, or 144-240 med.size hoes
2. Laterite?	Shaft w/2 tuyères	2.4 kg	3 1/2 mo.	35 kg, or 105–175 med.size hoes
3. ?	Bowl w/1 tuyère	4 kg? (12-14kg unrefined bloom)	2 mo.	32 kg, or 32–64 lg. size hoes
4. Laterite?	Shaft w/1 tuyère	4 kg? (10–20 hoes)	2 mo.	32 kg, or 80–160 med.size hoes
5. Laterite	Shaft w/2 tuyères	10 kg?	4 mo.	160 kg
6. Laterite	Shaft w/6 tuyères induced draft	8–30 kg	3 1/2 mo.	112–420 kg, or 56–210 very lg. hoes

* Smelts estimated to take about one week.

Geographical locations and sources:

1. Upper Ogowé, based on Georges Dupré, *Un Ordre et sa Destruction. Anthropologie des Nzabi* (Paris: ORSTOM, 1982), pp. 101–103.
2. Tchiungo-Ungo (lower Cokwe), based on Jose Redinha, *Campanha Etnografica ao Tchiboco* (Lisboa: CITA, 1953), p. 137.
3. North of Bululu (Aruwimi), based on George Hawker, *The Life of George Grenfell* (New York: Revell, 1909), p. 446.
4. Shaba, based on *Missionary Pioneering in Congo Forests,* ed. Max Moorhead (Preston, England: Seed & Sons, 1922), pp. 186, 195.
5. Ndop Plain (Cameroon), based on Jean-Pierre Warnier and Ian Fowler, "A Nineteenth Century Ruhr in Central Africa," *Africa* 49, 4 (1979), pp. 338–39.
6. Kasungu (Malawi), based on David Killick, *Technology in its Social Setting: Bloomery Iron-Smelting at Kasungu, Malawi, 1860–1940* (Ph.D. diss., Yale University, 1990), pp. 150–51.

These technical and environmental factors could indeed increase or decrease smelters' production volume, but more than anything it was labor organization that could make the greatest difference. Depending on the size of the labor force, a smelt could last a month or more, or it could take only a week. How smelters recruited and managed labor is the subject of discussion in the next section, and I argue there that as managers they had more than simple efficiency in mind. Nevertheless, for the purposes here of estimating the greatest potential number of smelts per furnace per season, I assume weekly smelts in Table 3-1, with the same furnace used and repaired throughout.

As Table 3-1 shows, the combined effects of specific types of furnace design, ore, and length of season could produce very different results. Pairing tuyères and using rich ore offset the short dry season and small scale of furnace in the upper Ogowé River. On the other hand, the type of ore used in Tchiungo-Ungo must have been of poor quality given the furnace's relatively larger scale and the longer smelting season. However, it must be kept in mind that these are isolated examples, and that smelters did not achieve precisely the same results over time. They do provide at least an indication of the possible range of output per furnace per season, which then can be compared to a rough estimate of yearly demand for iron from consumers.

Demand for iron implements came from many segments of society, one important one being farmworkers. There is at least some anecdotal evidence that when smelters in certain locales referred to projected furnace yields they did so in terms of the numbers of hoes or axes that they estimated would be produced. I have therefore translated my estimates of furnace yields per season from kg into hoes, although again, calculating the estimated number of hoes that could be made from a given amount of iron is not a simple matter. Agricultural tools were not uniform in the region, and so I have had to take into account various dimensions and weights for different types of hoes. Medium sized hoe blades weighing about 200–300 g were used in the western rainforest areas during the nineteenth century, as were small knives used in cultivation, which weighed less. Larger hoe blades weighing from 500 to 600 g and more were used in the east and in some areas of the southeastern savannas. Thus the type and size of hoe made is yet another factor that must be taken into account in calculating furnace output and especially, in calculating whether that output could meet the demand for iron tools in the agricultural sector of the economy.

The next step is estimating the number of hoes that were needed per year. Reliable population figures for central Africa are hard to come by for the present and are almost nonexistent for the past. My estimates of the yearly demand for iron agricultural tools are based on what is believed to have been the population density of the rainforests and of the savannas in about 1880.[18] I assume women were the major farmers, though males served in some places as slave labor, and so I estimate conservatively that half the population was involved in cultivating the fields. Dividing this figure by an estimated average lifetime of medium- and large-sized hoes,[19] I arrive at a figure for yearly iron demand. It appears that in the rainforests, where population densities were low, roughly four thousand furnaces would have had to operate throughout the dry season to provide enough iron for agricultural use alone. The southeastern savannas, an area half the size, would have needed over five thousand furnaces for its greater density of population.

These figures do not seem at first to pose a major problem in theory, but in practice this level of production would have been very difficult to maintain over time. And in practice, given the shifting patterns of settlement and their uneven densities, demand would have been experienced much more dramatically in specific places and at certain times, less so in others. Intensification of smelting at a particular site eventually would have led to the depletion of ore supplies, and in some cases high quality charcoal as well, problems which could be alleviated in the short term by trade.[20]

Buried in the literature are some scattered direct references to a trade in iron ore during the nineteenth century. One such example from Gabon states that the Tsengi subgroup of the Nzabi traded iron ore, bringing it from Ndendé to the Mouila district, a distance of about 75 km as the crow flies.[21] But other indirect references exist as well. When iron ore was transported from one place to another persistently according to a specific trading pattern, the group importing the ore distinguished it in their language from local ore, calling it by another name. Smelting vocabulary in the upper Ngiri area serves as an example of how new loan words for iron ore were added to languages in just this way, leaving traces of these past trading patterns. The iron ore they imported was from areas where Ngombe river traders operated, and as a consequence their borrowed term for iron ore, *makele*, comes from the Ngombe language.[22] Other such examples show that trade in iron ore was not unusual during the nineteenth century, especially along the river networks. In the area of the upper basin, Ngombe traders carried ore as far as 500 km, between the upper Zaire and the upper Salonga rivers.[23] Similar indicators of trade in iron ore show up in the languages of the middlemen merchants: there are at least four specific terms for iron ore in Ngombe;[24] and in Ifumu (Téké), there are two terms for iron ore, each one shared by other language groups within the Téké trading spheres.[25] Ore was traded overland as well, most prominently among ironworkers in Shaba and between them and as far as the Yao to the southeast.[26]

Charcoal was used in greater bulk quantities than ore in smelts, and so supplies of the specific hardwoods for making charcoal could become an especially acute problem. Blacksmiths in Sundi (Kongo) villages passed this problem on to their customers who had to supply their own charcoal if they wanted any smelting or forging done.[27] Another option was to substitute other species of trees for the most preferred hardwoods commonly used for making smelting charcoal. What the estimates in Table 3-1 suggest above all is that smelters could not sustain in one permanent place the levels of production year after year that would have been necessary to supply the demand for iron, especially as populations grew.[28]

So although ore and hardwood resources were generally available throughout the region, intense production levels could potentially lead to resource depletion in specific locales. If and when that happened, the portability of their equipment offered smelters a way out, for they could move to more richly endowed areas. Hence the interaction of iron production with changing environmental conditions encouraged trade of raw materials in some cases, relocation or itinerancy of workers in others. Sakata smelters are an example of the latter, traveling and smelting between the middle Zaire and Lake Mayi Ndombe during the nineteenth century, and then up to the middle Lukenie, leaving in their wake evidence of heavy forest clearance. In short, smelters' geographical mobility offered them not only safety and security, but also in some instances continuity of ore and especially fuel supplies.

Most importantly, however, there was never enough iron in its workable, metal form to meet the region's aggregate demand for it. Hence this scarcity of iron metal gave it an intrinsically high value, which goes a long way toward explaining why there was such a complex trade in iron goods and currencies that existed in nineteenth century central Africa. Although in the distant past iron undoubtedly was traded from production sites to places where iron was not produced at all, later trading patterns in the nineteenth century tell a different story. Centers where iron was produced for export were often also importers of iron and iron goods from elsewhere. In short, the pattern of distribution of smelting workshops, with bloomery furnaces so widely disseminated throughout the Zaire basin, should not be taken to mean that smelters were subsistence producers, making iron only for local use.[29] Patterns of trade in iron, described in the next chapters, demonstrate otherwise.

Production was complex, and so was consumption. Although strict quantification is not possible, a general survey of iron products and who used them shows that iron was not distributed equally across populations. For the poor, especially, it would have been important to guard and save even small morsels of iron, which helps to explain some of its ornamental and amuletic uses. Moreover, the slag inclusions in bloomery iron made it easy to heat weld and therefore suitable for recycling. But welding scrap together was not always necessary. Worn out hoes and hatchets, for example, still contained enough iron mass that they could be reforged to make smaller articles like arrowheads, cooking knives, razors, and pendants. The most massive iron products were ironworkers' tools, heavy swords and lance blades, and hatchets. Aside from the ironworkers themselves, it was free men, the wealthiest portion of the population, who were the greatest consumers of iron, for as warriors, hunters, and clearers of forests they used up greater numbers of these larger, weightier metal implements.[30] In contrast, their Twa clients consumed lesser amounts of iron. For although Twa hunters did use some iron lance and arrow blades, their most effective weapons were plain wooden arrow shafts, barbed to hold various kinds of deadly poisons.[31]

Most striking are the indications of gender inequalities in the consumption of iron. Crops and farming systems varied in the region, and women, who were often the primary farmworkers, did not always work with iron-bladed hoes. In areas north of the lower Zaire, for example, women worked with all-purpose iron knives,[32] as did women in the central basin. Such knives were smaller in scale and mass than hoes, requiring less metal to make them. There were also rather extensive areas in the middle and upper Zaire, lower Kasai, and in Gabon where digging and planting sticks were used in rainforest vegeculture. Hoes made out of wood were used in the upper Lukenie-Sankuru area, and also in at least one part of the savannas of the upper Kasai.[33] These were not economic backwaters, but were areas where smelters operated and several types of iron currencies circulated. Thus, contrary to the assumption that iron was produced and used mainly for local agricultural purposes, it could well have been the women cultivators of west central Africa who felt the scarcity of iron most frequently.[34]

DIVIDING AND MANAGING THE LABOR PROCESS

Labor requirements in smelting were variable, depending on the size of a furnace and especially on how many assistants ironworkers decided to recruit. Conflicts arose with demands for labor in food production, since smelting was usually confined to dry seasonal periods when ores could be mined and roasted or dried in the sun, and when wood could be charred to make the fuel. Dry seasons varied in length from two months or less in equatorial latitudes to up to four months or more in savannas beyond the fringes of the rainforests. Dry season was a time when male labor was also needed in clearing land for fields, as well as for collective hunting and fishing expeditions.[35] But food production was diverse enough, and iron products so necessary and profitable, that ironworkers did develop and practice specialized skills. As free men, they were also patriarchs, supported in large part by the work of women, who were the primary gatherers and farmworkers, and by the work of younger men and slaves. Free men in some areas were also supported by surpluses of wild game their Twa clients were willing to trade in exchange for farm or craft products. And where we have evidence for ironworking in fishing communities, where free men were most heavily involved in food production, we find that demand for iron could be sufficient to support full-time specialists.[36]

An iron smelt was not inherently labor intensive while the furnace was in operation, but the preparations leading up to it were in fact the major bottleneck. Some villages in the Aruwimi River area were described in the late nineteenth century as specializing in smelting, selling their iron to river traders. Ironworkers there overcame the labor bottleneck by recruiting what were called family members, most likely including slaves, to assist in the preparation for smelts. Together, these workers transported ore back to their villages, sorted and broke it up into smaller pieces, and spread it out to dry in the sun outside their homes.[37] This example and others like it suggest that the problem of labor shortage was not easily resolved. On one hand, a smelter who wanted to intensify the output of his smelting workshop could do so by investing in wives and slaves. On the other hand, by doing so he would be inviting potential conflicts within the household. For if a greater volume of production was to be achieved, it would be women, especially, who would bear an extra burden of processing ore on top of their already heavy work load of tending to food crops and household responsibilities.

Smelters focused their own energies on building the furnace, and in the Aruwimi case, they built several at a time on the outskirts of their villages. Preferred sites were at or near termite mounds, where processed clay was readily available for lining the inside of the furnaces, which were bowl types, about a meter long and half a meter deep. Bellows, made of wood and covered with antelope skin, could be used repeatedly, but the clay tuyère of the furnace had to be built anew for each smelt. The smelt proper reportedly lasted about four hours, producing a bloom weighing up to twenty five or thirty pounds which could be sold as it was or first refined to get rid of lumps of charcoal and unwanted slag. Even without the labor of refining, blooms were valuable. A British Baptist missionary, for example, bought an unrefined bloom along the Aruwimi in 1899 for four yards of strong calico.[38]

But labor was not divided simply to achieve economic efficiency. Scarcity of iron was exacerbated by the smelters themselves who limited their own numbers and the numbers of furnaces, thereby insuring that iron would remain a valuable commodity. Looking more closely at the labor process, one sees how critical the issue of management was in allocating tasks strategically so that the knowledge and skills in smelting remained under their strict control. Master smelters divided labor not only by the tasks necessary in carrying out the production process, but they also carefully divided labor by social group, in order that key knowledge remain esoteric and available only to certain selected men. This becomes apparent only by following examples of how smelters organized the smelting process itself and the preparations leading up to it.

One area where there is sufficient information about the organization of smelts lies inland from the left bank of the middle Zaire River. In this region, bordered by the Zaire River, the lower Kasai, and Lake Mayi Ndombe, many villages had been founded by ironworkers, and among these were smelting specialists.[39] The furnaces they built in the first years of this century were shaft furnaces about two and a half meters high, the walls reinforced by wrappings of vine. There was an open tunnel on one side of the furnace at its base, and the number of tuyère openings varies with different accounts from one to four.[40] The master smelter directed the operation, and since the bloom belonged to him, he apportioned it into units that were then paid as salary to workers. He sold the remainder to neighboring smiths who frequently traveled to smelting centers to purchase iron.

Joseph Maes, who visited this area in 1913–1914, was able to add more specific information and some historical depth to our knowledge of ironworking there.[41] Additional details about furnace construction and the smelting process come from Maes's interviews with Sakata elders and former smelting experts near the Fimi River, and probably also from the remains of old furnaces. He noted, for example, that the shaft furnaces differed in some of their features from one area to another, though his detailed description of the furnace construction process refers only to one type. It is unclear whether or not he himself witnessed a smelt first hand. Doubts are raised by certain anomalies in his account, and also by his claim that by the time of his visit, only the Sakata smelters along the upper Lukenie were still producing iron. Nevertheless, he can generally be relied upon for having based his research on the testimonies of knowledgeable witnesses to smelts.

The master smelter played a crucial managerial role in initiating and coordinating each iron smelt. He recruited family labor to prepare for constructing the furnace, taking advantage of their expertise in certain gendered tasks. Women, who were potters, prepared and kneaded the clay for the furnace walls, while men and boys cut the branches and vines that were to be used as supports. But the work involving knowledge specific to the functioning of the furnace and the selection of ore and fuel was in an entirely different category.

Mining and the making of charcoal were tasks apparently reserved for assistants chosen by the master smelter. Only men who were among those classed as "smelter-smiths" could take part in mining. This work was not particularly strenuous, as the iron ore was available in deposits close to the surface, no deeper than two meters down. The blocks of mineral were broken up into regular sizes by hammering with large hammers, and once the ore was sorted it was taken to the furnace site. Other smelter-smiths manufactured the charcoal fuel for the furnace. They and their designated assistants searched the forest, where the master charcoal maker chose the specific trees to be used and supervised the work. Hardwood resources were sparse in this area by the early twentieth century; the preferred trees were hard to come by, and it was necessary to travel for several days to find the right ones. Workers carried the wood to an isolated spot where they then roasted it slowly in an enclosed, temporary furnace, and when the process was completed, the charcoal was transported to the smelting site. The master charcoal maker Maes spoke with estimated that about three times as much charcoal as mineral ore was necessary per smelt.

The master smelter himself was in charge of constructing the furnace, and he began by choosing the precise spot where it would be built. He traced a circle in the ground where he then planted branches that would serve as the furnace armature. Around the branches he and his assistants wove a wickerwork structure in the form of a truncated cone, and against the walls of this structure they applied a first layer of clay. After a day of drying, another layer of clay was added to the structure, and this process continued until the master smelter judged the furnace walls to be sufficiently thick. Then he traced another circle in the ground, planted more branches, and created another wickerwork structure woven with vines. He applied clay at the intersecting points and worked the exterior clay wall to a smooth finish. The final exterior dimensions were three to five meters in diameter at the lower base, and two to two and a half meters diameter at the top, with the thickness of the furnace walls therefore greater at the bottom. Furnace height was about two and a half meters, sometimes up to three meters. Around this he built a shelter with a high roof which completely enclosed the furnace.

From this point onward, the master smelter's specialized knowledge was carefully controlled. All women, children, and men who were not smiths were sent back to the village, and prohibited from returning. The master smelter alone entered the furnace shelter and made tuyère slots and also the tunnel opening, all of which reached into the heart of the furnace. At the exterior of the furnace wall, the tunnel opening was about forty cm wide and twenty-five cm high, and along its passageway into the furnace was a long groove about ten to fifteen cm deep. The opening could be closed up with a clay cover. Although Maes does not identify it as such, it is apparently a feature that allowed for slag tapping to occur[42] and also doubled as an observation opening that could be used by the master smelter to view and assess the changing internal conditions of the furnace. Before the smelt could begin, the furnace interior had to be prepared. This, too, was a task only for the master smelter, who filled the furnace with dried wood and kept it burning for an entire day. The process served to bake the interior furnace wall, thus adding to its insulating ability.

Up to eleven men were involved in the actual smelt, and these men also were selected by the master smelter on the basis of their skill.[43] At least two were involved in charging the furnace with the proper proportions of ore and charcoal, while the master smelter observed activity in the working area of the furnace through the opening at the base. The first charges were apparently of charcoal only, and then the next charge consisted of a mixture of two parts charcoal to one part ore. Finally the fuel in the furnace was lit, and the bellows operators set to work in shifts of two men each. Henceforth the master smelter directed the timing and variable composition of the charges, which consisted of up to four parts charcoal to one part ore. He also instructed the bellows operators when to increase or decrease air flow. After about two days and two nights, the smelt was completed.

This description shows us just how strictly master smelters could control the entire process, but in some important respects it is not a representative case. These masters directed all aspects of the process of making iron, assuring their monopoly control over the full range of the necessary knowledge and skill in the preparations for the smelt and for the smelt proper. They were highly protective and secretive, enclosing the furnace completely within a walled structure, whereas in other areas of the Zaire basin smelting furnaces were sheltered only by roofs. Sengele elders further north, just to the east of Lake Mayi Ndombe, recalled how smelters (probably from Sakata communities) had traveled through their lands southward during the nineteenth century, building furnaces and smelting iron, selling it locally, and then destroying their furnaces before moving on.[44] They did not impart any aspect of their smelting expertise to Sengele smiths, and their furnaces reportedly were the only ones along the middle Zaire that had shafts and multiple tuyères.[45]

Other examples of smelting in the region provide evidence that such control was not always so strict, undoubtedly the result of masters selectively broadening their labor pool to intensify production. In some places, prominent Twa men had become the recognized experts in charcoal making, and in others it was women and children who were involved in transporting, preparing, and sorting the ore or making charcoal.[46] In other words, ironworkers as patriarchs could and did, when the need arose, recruit labor in various ways while preparing for smelts, from pools of family, client, or slave workers, taking care, however, to set limits on how much technical expertise these workers could acquire. In doing so, they partially overcame the production bottleneck, shortening the time it took to prepare for a smelt.

At the same time they fiercely controlled what mattered most, the furnace. That was where master smelters drew a line that few could cross, keeping for themselves the expertise in building a furnace and operating it. In accounts of smelting there is agreement on its central importance, for it was the key to making that precious commodity, the bloom of iron metal. All workers but a select few were kept away as the master directed the furnace construction and all aspects of its operation during the smelt itself. The furnace was his and his alone.

RITUAL, MANAGEMENT, AND THE TRANSFER OF TECHNOLOGY

An iron smelt was both a production process and a public performance, and its outcome could significantly enhance or possibly deflate a master smelter's reputation. For although smelting skills were guarded as secrets, they were deployed in highly audible and visible ways. Smelts were boisterous affairs and attracted a good deal of attention, as names of forebears or famous ironworkers of the past would be called out to aid the work in progress, and as bursts of song served to revivify tiring teams of bellows operators.[47] Smelts generated a lot more than iron.

Ritual played an important role on several levels. Ritualized precautions involving the entire community were believed necessary to insure that a smelt would succeed. During the smelt itself, in case after case throughout the region, women, children, and most men were prohibited from entering the immediate area. None of the participants were allowed to leave the premises, nor could any of them have contact with others in the village. Sexual relations with women were especially forbidden, and special arrangements had to be made for delivery of food, so that there would be no improper interference with the smelt. Ritual experts consecrated the furnace site, and proffered an array of "medicines" around and in the furnace itself, both before and during the smelt to ward off misfortune. The many variations of these prohibitions and precautions demonstrate how uncertain the process truly was—and how smelters had built into it many possible ways of deflecting blame for any failures.[48]

Ritualized prohibitions also served to insure that training and experience would be limited only to certain individual men. The entire process was organized hierarchically on the basis of skill, depending ultimately on the master smelter's technical and social expertise. While it has been stressed in the literature that women were excluded from acquiring this expertise, so, too, were most men. And here we arrive at the issue of how smelters' control over the process restricted the dissemination of smelting skills and, as a result, the number of furnaces in operation during a dry season. For although it has been shown that there were both technical and environmental constraints in the production of iron, it was the deliberately created scarcity of technological knowledge and skill that was the most important factor in limiting the numbers of furnaces at work.

It would be a mistake, however, to equate monopoly control with an entirely closed profession. Within a master's immediate locality, metallurgical training tended to follow family lines, but not strictly. If monopoly control of their skill had been completely maintained and confined to transmission within the family, that is, if smelters only trained certain of their sons, the historical patterns of change in smelting technology throughout the region would be relatively straightforward and easy to trace. But such is not the case, and instead, the combined evidence presents a very complex picture. By identifying certain specific techniques and equipment that were shared across ethnic groups, one can discern multiple instances of technology transfer from different directions and at different times. This picture, a kind of collage made up of discrete, overlapping, and intersecting elements, could not be the result of presumed group processes like "lineage segmentation" or long term language differentiation. Instead, it could only have come about by way of the contingent social relationships and training arrangements that ironworkers created among themselves.

Indeed, master smelters did have professional ties beyond their own workshops and communities. It was possible for well-connected blacksmiths from neighboring areas or other language groups to strike a deal, convincing a smelter to train him to smelt, and such deals were indeed struck in the past by, among others, Ekonda smiths and also the ironworkers of the upper Ngiri. In those oral accounts we have from master smelters, there is usually some reference to their having been taught by another ethnic group, though the precise conditions are not very clear. There may have been instances in the past when skilled smelters were captured and put to work by force, thereby transmitting experience in smelting to their enemies.[49] Yet another line of transmission was the ritual specialist who by repeatedly taking part in smelts also stood to acquire a certain amount of the masters' technical expertise. Master smelters asserted economic authority by withholding entry into their occupation at the local level, and they asserted social authority by granting limited access to it regionally.

Oral accounts that tell of smelting knowledge crossing ethnic boundaries are confirmed by material and linguistic evidence. Not only do furnace designs exhibit similarities and differences across space and time but so, too, do the lexical terms for them and for other aspects of the smelting process. In other words, there is no straightforward, consistent relation between language or ethnic group and the way smelters who spoke that language smelted. Some smelters used equipment that was similar to that used by smelters in neighboring ethnic groups; in other cases, not all smelters from the same ethnic group smelted with the same kind of equipment.

Although the specialized words for smelting furnace were usually not collected into Bantu language word lists and dictionaries, there is evidence in some cases of the contact and borrowing of knowledge that took place between blacksmiths and smelters from contiguous ethnic groups. Over much of the southeastern savannas, for example, one specialized word for smelting furnace (with the root form *-lungu*) was used by a number of ethnic groups who were prominent nineteenth-century ironworkers, and the unbroken distribution suggests recent borrowing.[50] Another major distribution occurs in the westerly savannas, where a word with a different Bantu root (*-tengo*) refers to a type of furnace used all over that area in the nineteenth century.[51] At least one language group shares both terms, creating an area of overlap between the eastern and western savannas. In these cases, master smelters carried either aspects of the furnace design itself or certain innovative techniques across language barriers or into new areas of settlement, transferring aspects of the technology to others.

The spatial dimension of how smelting changed was more complex when these kinds of transfers occurred more than once over time. Such was the case among ironworkers speaking the languages of Luba-Shaba, Bemba, and Tabwa. Each of these languages has at least three different specialized terms for smelting furnace, each term apparently referring to a furnace of a certain type and age. And in each language, the historical patterns of technology transfer that can be inferred have a different configuration. In other words, where the craft tradition was maintained over the long term within an ethnic group, smelting was not acquired only once but was learned and relearned or improved, older furnace features and techniques continuing in some places, new ones adopted in others. This linguistic evidence is supported by the material evidence of the different furnace designs that have been observed within certain ethnic groups. At least two types of furnaces were used by Songye smelters; Cewa smelters used one type of furnace for smelting, another for refining, and had formerly also known a tall shaft furnace design for smelting; and a group of Chisinga smelters using shaft furnaces spoke about earlier furnaces they had used that were bowl types.[52]

Transfer of technology among ironworkers of different villages and language groups is not so surprising once it is recognized that master ironworkers promoted themselves, developed reputations, and were sometimes famous. The most successful smelters were well known outside their own communities, and people were keenly aware of where smelting was done in their environs. Traders, too, carried along with their merchandise the names of ironworkers and news of their abilities and reputations, sometimes over long distances. The more widely traveled and enterprising merchants knew where smelting was intensifying, and linked these production centers with communities willing to sell their ore.

There were still other ways of getting news about ironworkers. Mines were not only resource areas but were also important meeting places where interested parties could discuss and confirm certain smelters' reputations. For while there is some evidence that ore deposits were monopolized by small groups of ironworkers or by certain ethnic groups, houses, or land chiefs, there is also ample evidence that access to them was negotiated, and that conditions of use rights could change over time. In the Aruwimi River area, smelting was done in villages some distance away from the iron mines,[53] which meant of course that assorted groups of ironworkers from many different workshops frequently congregated there during dry seasons. In Angola, a similar situation was described for the ironworking area around Ndulo, where skilled and unskilled workers converged at the "iron mountains" to mine for ore and take it back home to be smelted in their villages, sometimes many miles away.[54] The rich deposits of iron ore in the Mayoko area of the upper Ogowé (Gabon) were exploited in the past by miners and their labor recruits from Kota, Nzabi, and other ethnic groups.[55] Since iron ore was not a rarity, overt conflict over specific deposits was probably infrequent. It was knowledge about individual skilled workers, which one might gain at mines and other such meeting areas that offered the possibility of arranging direct access to that most scarce resource of all, the master smelter and his expertise.

Smelting workshops changed into meeting and market places when they were successful in producing a bloom. Even the secretive, itinerant Sakata smelters sold some of their iron to smiths outside their own communities. After all, a secret is not worth much if you hold it fast and no one knows you are keeping it. If iron smelters had been entirely unequivocal about releasing any of their expert knowledge, they would not have amassed and wielded the social power that they had. That there was hope for negotiation with them added value to their work and their reputations and gave them opportunities to establish relationships with ironworkers and others beyond their own villages and language groups.

Technology transfers from one workshop to another most certainly took place across family and language barriers, but only some portions of the most general patterns of transfer can be identified and followed. Expert smelters, when asked, will recount from whom they learned their technological knowledge, and their oral testimonies confirm that ironworking skills were not confined within ethnic boundaries. However, the testimonies couch this metallurgical history in generic terms, referring to entire ethnic groups rather than individual workshops or masters, to ironworking as a whole rather than to specific techniques, furnace features, or tools, and in the end they streamline the process into only one single line of transfer. Lalya-Ngolu smelters, for example, told of learning iron metallurgy from the Mbole; Nzabi smelters learned from the Téké-Tsaayi; north Mbala smelters learned from the Hungana; and south Mbala smelters learned from the Songo.[56]

Testing these oral testimonies against other evidence, such as descriptions of furnace designs and certain lexical terms, raises still more questions. Comparing the furnaces used by Nzabi smelters with ones formerly used by their putative teachers, the general similarities in the equipment and smelting process corroborate the smelters' claims. However, the lexical term for smelting furnace in Nzabi is completely different from the Téké-Tsaayi term.[57] Plausible explanations for this are that the Nzabi smelters coined their own new term for the furnace, or that they already had one in use. In the latter case, it is possible that some Nzabi ironworkers already knew how to smelt but then learned additional or improved techniques from their Téké counterparts, choosing to remember only this more recent change. The presence of specialized loan words in a language can indicate technology transfer, but as this example shows, their absence does not then mean that no transfers ever took place.

Bringing together smelters' oral testimonies with other types of evidence can be very useful for showing how it is that the patterns of ironworking history can be so complex and why even simple relative chronologies are so difficult to construct. Cokwe ironworkers, who were particularly famous in the nineteenth century for both their smelting and smithing, will serve here as an illustration. When interviewed in the mid-twentieth century, individual Cokwe smelters told similar histories of their expert metallurgical knowledge. In one account, smelters in the upper Cokwe area not far from Dundo (Angola) stated: *mulimo wa lutengo wa Tupende ku nyima yalongesa Basoso, Basoso yalongesa Tulunda, Tulunda yangolesa Tucokwe* (the work of 'lutengo' originated in Pende country, Pende taught it to Soso, Soso taught it to Lunda,

Lunda taught the Cokwe).[58] Another account is more abbreviated, stating that iron metallurgy passed from the Soso to the Pende, and then to Cokwe ironworkers.[59]

What is not immediately apparent but is most interesting about this sequence is what it suggests about discontinuities in the history of ironworking within ethnic groups. Pende ironworkers, credited in the more reliable account with the development of the shaft furnace *lutengo*, were not smelters in the nineteenth century, and so must have stopped producing iron perhaps as long ago as several centuries. During the first half of the seventeenth century, forerunners of nineteenth-century Pende abandoned their former homelands along the middle Kwango River (Angola), an area known at that time for its iron mining and smelting, in order to escape the raids and political upheavals of Imbangala expansion. Some of these refugees were probably among those captured and sold into slavery, while some were absorbed, forcibly or by choice, into other communities or the territories claimed by newly forming bands of warlords. Many of them resettled far to the northeast where they formed independent chiefdoms along the right bank of the middle Kwilu River and where they created and maintained a Pende group identity.[60] In this new location, Pende ironworkers continued to operate their smithing workshops, but smelting was no longer practiced. Instead, they relied on the iron produced by Hungana smelters, who were the main specialists in the lower and middle Kwilu, and by other smelting workshops in Kwese, Mbala, Pindi, and Mbunda communities.[61] What happened to smelters and smelting technology among Pende forebears is not known, but as a result, there are no furnace or smelting descriptions and scant linguistic evidence that could be used to test the claim that Pende ironworkers developed smelting techniques associated with the *lutengo* shaft furnace.

If the oral account is accurate, and there were at one time ancestral Pende smelting workshops, that does make it possible, then, to suggest temporal brackets for the transfer sequence. It is more likely that the smelters passed their expertise on to neighboring Soso ironworkers before they left the middle Kwango, that is, before 1650, rather than after. And the earliest recorded observation of Cokwe smelting was made by Magyar around 1850.[62] Hence this particular sequence involving the transfer of smelting techniques between ironworkers in four different ethnic groups would have occurred over a period of at least two hundred years. Given that length of time and the training of smelters from various workshops that went on within these groups, it is unlikely that the original (pre-seventeenth century) furnace design and smelting techniques would closely resemble the ones used by Cokwe smelters in the nineteenth and early twentieth centuries. This problem will remain unresolved until teams of archaeologists can survey and excavate smelting sites in the middle Kwango area.

For more recent times, there do exist descriptions, drawings, and photographs of the shaft furnace called *lutengo*. All eight examples are from the first half of the twentieth century. One was used by Lunda smelters; six others by Cokwe smelters; and one was located in Lwena country. The furnaces were called *lutengo* in each of these languages, and the ones described show remarkably consistent structural features. Their forms were of similar construction, made of termite earth for permanence and insulation, and air flow entered the furnace through a single tuyère. The only noticeable difference was the absence of female imagery (frequently a feature of Cokwe furnaces) applied to the furnace surface in the Lunda example.[63] But it is still unclear how far back in time this particular furnace design goes.

Linguistic evidence hints that the shaft furnace *lutengo* might have originated along the middle Kwango sometime after the mid-seventeenth century. Although recorded vocabularies for smelting equipment and processes are woefully incomplete, those terms that we do have suggest a pivotal role for Lunda ironworkers. The language of the nuclear Lunda, uRuund, shows two different verbs meaning "to smelt": *-é:ngúl*, and *-rîté:ng*. The latter could have served as the stem for *lutengo* (although there is a problem with the tone), while the former can be linked to one of the northern Mbundu verbs meaning "to smelt," *-zenguluka*. This would take us back to Soso, which is a Mbundu-north subgroup, though we still lack data for Pende.

I present all of these details not to confuse the reader but to demonstrate the fallacy of expecting a simple coherent narrative history of ironworking. What is abundantly clear is the fragmentary nature of the data at present, and how a single source of evidence provides only the initial introduction to a very complicated set of stories. The transfer patterns described in oral traditions should be considered only as basic markers, for it is likely that there were multiple contacts over time, among different workshops, rather than the single one described. Ironworkers retain important memories, though these are highly reductive, referring to the most recent or general trends and to what were thought to be the most significant links in the chains of knowledge transfers that took place.

Above all, the patterns are complicated because ironworkers themselves had complex identities. They belonged to an occupational group that cut across family and ethnic ties, a craft tradition that they themselves were in the process of changing. Smelters, for example, adapted to new conditions, created new techniques, and added to their knowledge by retraining with neighboring or distant masters. Ironworkers also belonged to communities. As political and social conditions changed over time, whole groups of people might be forced to move elsewhere, either to protect their communities or to seek more land. Smelters moving along with their villages might or might not find suitable sources of ore or access to them.

And smelters who continued to work met new neighbors with whom they might or might not exchange technical information or experience. Mobility of ironworkers and of whole groups, combined with potential interethnic transfer of technology, lends the history of central African ironworking an almost unfathomable complexity.

Smelters have been called "masters of fire." It is more accurate to call them "masters of the furnace," or better yet, "fathers of the furnace," for their position as patriarchs provided the foundation on which their craft tradition developed. It was this position and its prerogatives that they exploited in managing and monitoring the raw materials, equipment, and conditions that brought about the creation of iron blooms over and over again. Reading the temperature of fire by its color is an essential skill of the successful ironworker, but it is not the only one, for a master smelter invested other aspects of his expertise just to reach those times in a smelt when temperature was crucial. Mastery of the technological process was important but it was not his only forté. Continuing and modifying the knowledge of furnace construction and the successful operation of a furnace was significant also in producing a host of social relations. As the furnace owner, a master claimed the authority to divide and mobilize the labor of women, children, clients, and slaves in preparing for a smelt, and to distribute the valuable bloom to consumers and merchants afterward. Equally important was how he managed skilled labor during the smelt, and embellished it with ceremony. For this was how he maintained his craft tradition while at the same time creating a highly prized and useful metal.

NOTES

[1] Lumbwe Mudindaambi, *Dictionnaire Mbala-Français*, 4 Vols. (Bandundu: CEEBA, 1977–81).

[2] Each example is of an established workshop with skilled labor operating the furnace, all of which necessitates training and organization built up over time, hence my inference that there was earlier, unrecorded evidence.

[3] For example, Mongo ironworkers preferred charcoal made from *Uapaca Guineensis*. Gustave Hulstaert, *Les Mongo. Aperçu Général* (Tervuren: MRAC, 1961). Lega smelters distinguished between the trees *muntungulu* and *kantengentenge*, that produced "light" charcoal and the trees *ngale, muzombo, kyume*, and *musase*, that produced "heavy" charcoal. Kyankenge Masandi Kita, "La Technique Traditionelle de la Métallurgie du fer chez les Balega de Pangi (Zaire)," *Muntu* 3 (1985), p. 92.

[4] Cyril Stanley Smith, "Metallurgy in the Seventeenth and Eighteenth Centuries" in *Technology in Western Civilization*, eds. Melvin Kransberg and Carroll Pursell (New York: Oxford University Press, 1967), Vol. 1.

[5] Here I define urban center as a distribution center for goods, either in trading or tribute networks or both. Large population may or may not be a factor. For a discussion of issues regarding urbanization in world history, see Witold Rybczynski, "The Mystery of Cities," *New York Review of Books* (15 July 1993).

[6] Letter from G. Grenfell, Nov. 1899, quoted in Hawker, *Life of Grenfell*, pp. 446–47.

[7] From the 1919 diary of W. F. P. Burton, published in *Missionary Pioneering in Congo Forests*, Max Moorhead, ed. (Preston, England: Seed and Sons, 1922), p. 166.

[8] It is not clear exactly who these smelters were, but they were not in the immediate environs of the new villages.

[9] See Mumbanza mwa Bawele na Nyabakombi Ensobato, "Les Forgerons de la Ngiri. Une Elite Artisanale parmi les Pêcheurs," *Enquêtes et Documents d'Histoire Africaine* 4 (1989).

[10] See, for example, Léon Guiral, *Le Congo Français* (Paris, 1889), pp. 161–70. For the Ikelemba, see H. H. Johnston, *George Grenfell and the Congo*, 2 Vols. (London: Hutchinson, 1908), Vol. I, p. 117; Camille Coquilhat, *Sur le Haut-Congo* (Paris: Lebègue, 1888), pp. 150, 158.

[11] A. Engels, *Les Wangata. Etude Ethnographique* (Brussels: Vromant, 1912), pp. 31–32; G. Hulstaert, "Aux Origines de Mbandaka" in *Histoire Ancienne de Mbandaka*, D. Vangroenweghe, G. Hulstaert, and L. Lufungula, eds. (Mbandaka: Centre Aequatoria, 1986), pp. 119–21. One of the furnaces was dated 1885 ad +/- 50, calibrated to 1680–1800 AD. See Eggert, "The Current State of Archaeological Research," p. 41; de Maret, "Recent Archaeological Research," p. 137.

[12] The smelting center may have involved a number of different groups of smelters; the group referred to here were from Sakanyi communities. Robert Harms, *River of Wealth, River of Sorrow*, p. 68; F. Bolesse, "Essai historique sur les Lusankani," *Aequatoria* 23 (1960), pp. 100–102, 105–106.

[13] For a discussion of some aspects of furnace design, see S. Terry Childs, "Style, Technology, and Iron Smelting Furnaces in Bantu-Speaking Africa," *Journal of Anthropological Archaeology* 10 (1991). And for a useful comparison of bloomery smelting in Africa and North America, see Gordon and Killick, "Adaptation of Technology to Culture and Environment."

[14] Personal communication, J. E. Rehder, Toronto, 27 September 1989. See also the hypothetical comparison by Sutton of two types of furnaces in Tanzania, showing how aspects of furnace design might be closely related to the type of ore used. High shaft furnaces using limonite ore with low iron content would require large amounts of ore and charcoal; if natural draft was used, smelts would take a relatively long time (3 to 5 days) but would require little manual labor to produce the air supply. In contrast, a smaller furnace would require less ore and charcoal if the ore had a relatively high iron content; forced draft would mean more labor but the bloom would be produced more quickly, say in one day rather than three or four. And, based on hearsay, the smaller furnace apparently produced a larger bloom than did the shaft furnace. J. E. G. Sutton, "Temporal and Spatial Variability in African Iron Furnaces," in Haaland and Shinnie, *African Ironworking,* pp. 174–75.

[15] Here a note of caution must be inserted, for there have been attempts to ascribe to African ironworkers types of knowledge and expertise that they may not have had. Here the modifications in tuyères and shafts, which produced larger working areas in the furnace, may have increased the observed success rate of the bloomery process, rather than producing consistently larger blooms. It must be remembered that the workers were operating by memory and observation and not necessarily according to the procedures of formal scientific experimentation.

[16] This will come as no surprise to those who are familiar with the study of intensified smelting in the Ndop Plain of Cameroon. See Jean-Pierre Warnier and Ian Fowler, "A Nineteenth Century Ruhr in Central Africa," *Africa* 49, 4 (1979).

[17] I would like to thank David Killick for bringing these factors to my attention.

[18] Based on Jan Vansina, "Peoples of the Forest," pp. 79–80 and map on p. 110.

[19] Four years and six years, respectively. Although there exist other estimates for a shorter lifetime of a hoe blade in equatorial Africa, I have used the most conservative figures available here, based mainly on the averages noted on the Ndop Plain in Cameroon. See Warnier and Fowler, p. 338. Recently-made hoes may wear out in one season, however. See Ian G. Robinson, "Hoes and Metal Templates in Northern Cameroon," in *An African Commitment*, eds. Judy Sterner and Nicholas David (Calgary, Alberta: University of Calgary Press, 1992).

[20] I have found references to trade in ore overland and by boat, but none yet on trade specifically in charcoal, though it must have taken place, at least locally.

[21] Hubert Deschamps, *Traditions Orales et Archives au Gabon* (Paris: Berger-Levrault, 1962), pp. 27–28.

[22] For the Ngiri term, see Mumbanza mwa Bawele, "Les Forgerons de la Ngiri," p. 118.

[23] The Mbole of the upper Salonga River use the Lomongo term for ore (with the root form *-boko* or *-oko*) and also the Ngombe term *-kele*. This suggestion of ties with Ngombe traders is corroborated by the manufacture by Mbole smiths of the *ngolo* knife associated with workshops within the Ngombe trading sphere in the late nineteenth and early twentieth centuries.

[24] They are: *mangelo; libángá; likéle; mojegelé.*

[25] *Mbolo* (C languages) and *kele*—pointing to the northern basin (Ngombe) and also Mbole of upper Salonga (also C languages). Téké is a B language.

[26] Terms for iron ore with the root *-tapo* are shared among the languages of Bemba (M42), Tabwa (M41), Lala (Lamba) (M52), Kaonde (L41), Nyanja (N31), and Yao (P21), all contiguous language groups in recent times.

[27] Karl Laman, *The Kongo*, 4 Vols. (Uppsala: Studia Ethnographica, 1953–68), Vol. I, pp. 122–24.

[28] On the problem of deforestation in reference to iron smelting, it should be reiterated that deforestation is caused generally by overpopulation and land clearance, not specifically or solely iron smelting *per se*, although of course the two are closely related. On the ecological effects of very intense iron smelting in one West African case, see Candice Goucher, "Iron is Iron 'til it is Rust: Trade and Ecology in the Decline of West African Iron Smelting," *JAH* 22 (1981); and L. M. Pole, "Decline or Survival? Iron Production in West Africa from the Seventeenth to the Twentieth Centuries," *JAH* 23 (1982).

[29] See also the impressive evidence for smelting on the Téké plateaus in the sixteenth century, interpreted convincingly as far from subsistence oriented. Dupré and Pinçon, *Métallurgie et Politique,* pp. 174–75.

[30] The logical inferences I am making here are based on known technical qualities of the iron, my argument and conclusions about its high value, and my data and observations from handling thousands of ironworkers' products.

[31] Non-Twa hunters also used such arrows. See, for example, the Kete example in Hildegard Klein, ed. *Leo Frobenius: Ethnographische Notizen aus den Jahren 1905 und 1906*, 4 Vols. (Wiesbaden: Steiner, 1985–90), Vol. II, p. 202, # 376.

[32] Personal communication, Marie-Claude Dupré, 9 March 1993. I would also like to thank her for sharing with me excerpts from her unpublished manuscript, "L'outil agricole des essartages forestiers; le couteau de culture au Gabon et au Congo," (1990).

[33] Hermann Baumann, "Zur Morphologie des afrikanischen Ackergerätes," *Koloniale Völkerkunde. Wiener Beiträge zur Kulturgeschichte und Linguistik* 6, 1 (1944), pp. 207, 210, 231, 233, 253–56. The specimens Baumann illustrates date to the late nineteenth and early twentieth centuries. See also Klein, *Frobenius*, Vol. II, p. 192, # 313.

[34] A Mongo proverb contrasts the relatively small women's field knife to the heavier, stronger iron implements used by men, stating that even though the former is insufficient for certain tasks, sometimes one just has to "make do" with it. Roughly translated, it describes *ikoko*, the knife used by women in working the fields, as a knife to be used only as a last resort for butchering an antelope or killing a snake. Gustave Hulstaert, *Proverbes Mongo* (Tervuren: Musée du Congo Belge, 1958), p. 119.

[35] See Vansina, *Paths,* pp. 40, 83, 86, 90–91.

[36] See Mumbanza mwa Bawele, "Les Forgerons de la Ngiri," on Ngiri communities.

[37] P. Nahan, "Reconnaissance de Banalia vers Buta et retour à Bolulu," *Belgique Coloniale* 4 (1898), pp. 557–58.

[38] Nahan, pp. 557–58; Hawker, *Life of Grenfell*, pp. 446–47.

[39] M. Baeyens, "Les Lesa," *La Revue Congolaise* 4 (1913–14). His name was misspelled in Joseph Maes' published paper, and Maes mistakenly listed the date of his journal entry as 1919 instead of 1909. J. Maes, "La Métallurgie chez les populations de Lac Léopold II-Lukenie," *Ethnologica* 4 (1930), pp. 69–70. The people he was describing were speakers of Sakata.

[40] Baeyens published a drawing that does not show tuyère openings; Maes' article includes a drawing that shows four sets of four-drum bellows; a later published drawing shows one tuyère opening for a four-drum bellows. Maes, "La Métallurgie," fig. 1; Foquet-Vanderkerken, "Les Populations indigènes des territoires de Kutu et de Nseontin," *Congo. Revue Générale de la Colonie Belge* 5, 2 (1924), pp. 163–64.

[41] Published in Maes, "La Métallurgie"; additional information in unpublished MS "métallurgie" and his journal de route, Papiers Maes, Archives Africaines, D 61, (3858) No. 68 and (3853) No. 9. Great care must be taken in using Maes' material, however, as he combined his own firsthand observations with reports he plagiarized from others. One such example occurs at the bottom of p. 4 in his unpublished MS, where he jumps from his account of furnace construction that he himself observed in the Lake Mayi Ndombe district to a word-for-word plagiarism of Nahan's description of Boa furnaces, which he does not cite.

[42] Baeyens mentioned men on the smelting team who would crush 'puddings' [sic] that were too large, which probably refers to slag that was running out of the tunnel during the smelt. Maes claimed that the groove was a crucible into which flowed the molten metal, but it is more likely that what flowed into the groove was molten slag.

[43] Baeyens describes a smelt as involving eleven men; Maes does not provide a total. Certainly the labor needs were variable according to the judgement of the master smelter each time.

[44] Maes, "La Métallurgie," pp. 68–69.

[45] So-called Ngombe shaft furnaces further north are the closest examples, and there is no lexical evidence that would indicate that the two types of furnaces or smelting operations are historically related.

[46] Kriger, field notes, Lopanzo, 1989; Célis, "Fondeurs et Forgerons Ekonda"; Frank Read, "Iron Smelting and Native Blacksmithing in Ondulu Country, Southeast Angola," *African Affairs* 2, 5 (1902–3); Carlos Estermann, *The Ethnography of Southwestern Angola* (New York: Africana Publishing Co., 1976), Vol. I; Guiral, pp. 161–70; Mumbanza mwa Bawele, "Les Forgerons de la Ngiri," p. 117; and Laman, *The Kongo*, Vol. I, p. 48.

[47] Burton remarked on the noise of Songye smelters working at 4 a.m. shouting for hoes and axes to come out of the red iron ore. From the 1919 diary of Burton, published in Moorhead, *Missionary Pioneering*, p. 186. See also Piet Korse, "La Forge," *Annales Aequatoria* 9 (1988), pp. 30–31; and Kriger, field notes, Lopanzo, 1989.

[48] For detailed descriptions and analyses of such prohibitions, see Killick, *Technology in its Social Setting*, pp. 123–39; and Herbert, *Iron, Gender, and Power.*

[49] A Kongo account alludes to such a possibility, involving supposed Vili captives of the Sundi. Laman, *The Kongo*, Vol. I, p. 18. Though it is unlikely to have happened in fact, it is important evidence of what a Kongo man believed to be an avenue of technology transfer among smelters.

[50] Words for furnace sharing this root form are found in Lwena (K14), Songye (L23), Luba-Shaba (L33), Hemba (L34), Tabwa (M41), Bemba (M42), and Fipa (F13). The tall, induced draft shaft furnaces known by this term were not as widely distributed as the lexical term itself. Some Bemba oral testimonies referred to early inhabitants of Bembaland as *bashimalungu*, or 'ironsmelters.' Lungu is also an ethnic term; the Lungu were among the several neighboring groups who supplied iron to Bemba communities in the nineteenth century, and some Lungu ironworkers immigrated into Bemba chiefdoms, paying tribute in hoes and axes. Andrew Roberts, *A History of the Bemba* (Madison: University of Wisconsin Press, 1973), pp. 75, 184–86.

[51] The shared term for furnace with the root *-téngo* occurs in Lwena (K14), Cokwe (K11), and Lunda (K). However, in this case, the widespread distribution of it was probably the result of Cokwe expansion in the nineteenth century. Interestingly, there are two different root forms for the verb "to smelt" in uRuund (K23), the one here shared with other K languages, and another one shared with H languages.

[52] For Songye furnaces, see the bowl furnace described by Torday in Emil Torday and T. A. Joyce, *Notes Ethnographiques sur les Populations habitant les Bassins du Kasai et du Kwango Oriental* (Tervuren: Musée du Congo Belge, 1922), p. 38; and the shaft furnace described by Burton in Moorhead, *Missionary Pioneering,* p. 186. For Cewa furnaces, see D. W. Phillipson, "Cewa, Leya, and Lala Iron-Smelting Furnaces," *South African Archaeological Bulletin* 23 (1968). And for Chisinga furnaces, see W. V. Brelsford, "Rituals and Medicines of Chishinga Ironworkers," *Man* 49 (1949).

[53] Nahan, "Reconnaissance de Banalia," pp. 557–58.

[54] Read, "Iron Smelting and Native Blacksmithing," pp. 44–49.

[55] G. Collomb, "Quelques aspects techniques de la forge dans le bassin de l'Ogooué (Gabon)," *Anthropos* 76 (1981), p. 50.

[56] See F. de Ryck, *Les Lalia-Ngolu* (Anvers: Le Trait d'Union, 1937); G. Dupré, *Un Ordre et sa Destruction*, p. 36; and Torday and Joyce, *Notes Ethnographiques* (1922), pp. 349–50.

[57] Both were bowl furnaces with two tuyères, smelts lasting about one day. The Nzabi term for smelting furnace is *mbimba*, while the Téké-Tsaayi term has the root *-nzuru*. G. Dupré, *Un Ordre et sa Destruction,* pp. 100–102; and M.-C. Dupré, "Pour une Histoire des Productions," p. 213.

[58] Marie-Louise Bastin, "Le Haut-Fourneau 'lutengo': Opération de la Fonte du Fer et Rituel chez les Tshokwe du Nord de la Lunda (Angola)" in *In Memoriam António Jorge Diaz* (Lisboa: Instituto de alta cultura junta de investgações científicas do ultramar, 1974), p. 68.

[59] Augusto Guilhermo Mesquitela Lima, "Le Fer in Angola," *Cahiers d'Etudes Africaines* 17, 2-3 (1966–67), p. 350. Lima does not name his Cokwe sources, and does not state whether they were ironworking experts or not; in other words it is not clear whether this is a popular historical account, or whether it is more reliable than that. Bastin's sources were indeed reliable.

[60] Miller, *Kings and Kinsmen*, pp. 42, 70–73, 109–111; Jan Vansina, *Kingdoms of the Savanna* (Madison: University of Wisconsin Press, 1968), pp. 93–97, fn 44 p. 278.

[61] See Léon de Sousberghe, "Forgerons et Fondeurs de Fer chez les Ba-Pende et leurs Voisins," *Zaïre* 9, 1 (1955). The author convincingly disputes the statement by Torday that Pende ironworkers knew how to smelt, at least in their new location.

[62] N. de Kun, "La Vie et la Voyage de Ladislas Magyar dans l'intérieur du Congo en 1850–1852," *Bulletin des Séances de l'Académie Royal des Sciences d'Outre-mer* (Brussels) 6 (1960).

[63] Jose Redinha, *Campanha Etnografica ao Tchiboco* (Lisboa: CITA, 1953), pp. 103, 125, 129–40; Hermann Baumann, *Lunda* (Berlin, Würfel, 1935), pp. 80–84, Plates 23-1 and 24-1; and Bastin, "Le Haut-Fourneau."

4

Making and Changing Money: Iron Currencies

Botúli ntemáká lifelámá.
The blacksmith does not make anything that is unsuitable.
Botúli aowá, mbolo yélengana.
When the blacksmith dies, the iron goes too.
(When your benefactor is gone, you are in need.)

—Mongo (Basankusu)[1]

While the previous chapter focused on smelters, the primary manufacturers of iron metal, this one examines work at intermediate stages of manufacturing, namely the processes of refining blooms and making semifinished products. Ironworkers who mastered these processes were literally making money, for it was in a variety of basic, generic forms that iron circulated and performed as currency units throughout the region. The chapter begins with the fundamental question of the value of iron as a metal. Building on the previous chapter, where it was shown that iron was scarce, I present further evidence that there was a much higher value attached to iron in central Africa than would be the case in an industrialized economy. Much of the discussion is taken up with identifying major iron currency forms and delineating the geographical zones in which they circulated. My examples are units that I have identified as having been current in the late nineteenth century, basing my judgement on a combination of museum accession data, lexical terms, and patterns of geographical distribution for the units. There is no single chronological scheme here for the histories of the currencies themselves. In some cases, for example, I suggest that a currency form had been established for a longer period of time or had been partly displaced by others. I also trace whenever possible the ways iron could move in and out of commodity and currency networks. The chapter ends with several examples of how small and large denomination currency units worked together.

Altogether, my evidence demonstrates new and compelling reasons why the blacksmith in particular was esteemed and revered, for he was both the maker and the converter of iron currency, a key aspect of his work in precolonial times that up until now has not been adequately acknowledged or explored. How he worked, where, and with whom affected the supply, flow, velocity, and conversion of money. This offers yet another reason for the "mystique" surrounding the blacksmith, and his acknowledged social power and prestige.[2] Based on this new evidence, I argue that iron currencies linked rural areas with the arteries of external trade networks, helping to integrate regional economies. I also demonstrate that the specific forms of the currency units were semifinished metal goods, temporarily frozen at a specific intermediary stage of manufacture. The visual evidence that manufacturing processes were integrated into the monetary forms suggests very strongly that it was smiths who were in control of these regional commodity currency systems.[3]

VALUES OF IRON

In present-day western terms, iron is a cheap and very practical metal, obviously not a precious one. It owes this assessment to its association in peoples' minds with mass-manufactured goods, heaps of them rusting and deteriorating in scrap metal yards, and a modern history that is very much intertwined with industrialization. Other metals such as gold, silver, or even copper are thought of differently. They seem to occupy a separate, more exalted realm, being considered more pure or more intrinsically valuable, all three having been used in the past as coinage or for creating extravagant luxury objects or both. World market prices for metals at present show silver valued at sixty times the equivalent weight of copper, and copper at thirteen times the equivalent weight of pig or wrought iron from the factory.[4] Industrial-scale blast furnace production has turned iron into a very cheap metal indeed.

Such a striking contrast between the value of iron and that of other metals should not be taken as a normative or constant standard, for that would obscure the place of iron in other societies and at other times in world history. Iron was definitely a very practical metal in west central Africa, and at the same time it was valued as a semiprecious one, taking the form of useful commodities as well as extravagant luxury goods that will be discussed in the next chapter. It also circulated as bar iron and other currency forms that were used for purchasing goods and titles, paying fees and fines, and making bridewealth transactions. In some cases it even served as a unit of account. It was certainly valuable, but not necessarily for the same reasons other metals were.

While it cannot be denied that iron was valued highly in nineteenth-century west central Africa, why this was so is not easy to explain. That it could have a potentially high value comes at least in part from a combination of factors having to do with its material qualities and its peculiar kind of scarcity. First, it was bloomery iron, with great tensile strength and a relatively low carbon content,[5] a material that could be handily worked by the blacksmith into all sorts of products. Even more importantly, the slag inclusions remaining in bloomery iron gave it fusibility, so pieces of it could be pressure welded together without any additional fluxing agent.[6] As a result, iron products could be easily reworked, sharpened, rehardened, and repaired, and it was probably these qualities of malleability that central African blacksmiths had in mind when they referred to locally produced iron as "stronger" than iron from overseas. Second, these material properties rendered iron useful for making an extensive range of tools, weapons, and implements, and so the consumer demand for it was undoubtedly strong and steady. Markets for iron goods, from basic commodities to luxuries, existed at all economic levels and consumers spanned the social spectrum. Finally, as was shown in the previous chapter, there was limited supply, or scarcity. Demand could not be entirely met by bloomery furnaces even when iron production was intense, a constraint exacerbated by smelters who, by controlling their esoteric knowledge, limited the number of smelting teams that were active at any given time. And the smelting process was difficult and risky, making production costly and the iron valuable.

Technological and environmental limitations in smelting, unequal distribution of furnaces and their products, and above all, canny ironworkers, together created scarcities of iron in central Africa that were very different in kind from scarcities of, say, copper. Since far fewer deposits of copper ore exist in nature, the rarity of copper is more obviously apparent and its high value more immediately understandable. But the potential consumer demand for copper was not as great as that for iron. Iron surpasses copper and copper alloys in terms of its potential use value for a much wider array of products. In contrast, one finds copper serving most often as currencies, stored wealth, and luxuries. Such was the case in the Upemba depression burials, where as copper became more available it seems to have predominated in the category of body ornaments, but was turned into fewer utilitarian objects. The relatively greater frequency of iron ore deposits and the more widespread smelting of iron were thus offset by its uses in many more social and economic domains. In relation to demand it became a valuable metal, but its high value was based in large part on human assessments and actions, not just on environmental conditions.

Indications are that central African traders and consumers valued iron about as highly as they did copper. Some copper to iron equivalencies are included among standards of value in trade listed for the late nineteenth or early twentieth century. In each instance, by examining and weighing named currency units, I have been able to determine that a given amount of copper metal was roughly equal in value to that same amount of iron. In the north along the Ubangi River basin, heavy knife blades weighing an average of 330 g could be purchased with thirty *mitako* units, the equivalent of about 300 g of imported brass.[7] Far to the south and east of there, between the middle Lualaba and upper Lukenie, four iron spear blades called *ikonga* were worth one cast copper cross ingot from Shaba, each side of the equation consisting of about the same amount of metal.[8] In the upper Lukenie basin, trade data showed similar values between imported copper and iron. One *boloko*, a U-shaped unit of copper bar from overseas weighing from 800 to 1100 g, was worth ten units of *kundja* blades, or roughly 800 to 1600 g of iron.[9] These figures are not meant to imply that metal currencies were measured and assessed specifically by weight, though perhaps they were in some places. It was the mass that truly mattered to blacksmiths, and here I have estimated the mass of these variously shaped objects by weighing them. What they show is nothing like the disparity between the value of copper and the value of iron that one sees in the published prices of metals in the world market today. And so, given all of the factors mentioned above, it is not surprising that iron was turned into so many different currency forms.

COMMERCE AND CURRENCIES

The many iron currencies of the Zaire basin have hardly been noticed by scholars,[10] and the various and overlapping zones in which certain forms of iron currency units circulated have never before been mapped out. One reason for this is the emphasis placed on foreign trade in the written sources. Descriptions of long distance overland caravans and river networks in the region stress the external features of commerce, namely the imported manufactures from Europe, especially textiles, firearms, and metalwares that were used to finance the trade itself and the slaves and ivory that were transported to the coast for export. Central African economies used what are called commodity currencies,[11] i.e., products that were used as units of account and as means of making payments but that were not backed currencies or actual coinage. These forms of currency are sometimes difficult to separate from ordinary trade in commodities and, of course, they also changed over time. They were used in both the external and regional trading networks, but only a few of the regional currency forms were written about by outsiders. By the 1890s, the major currencies that were recorded by European observers included tokens in the form of beads and cowries, and commodities such as raffia cloth, all sorts of imported textiles, copper and copper alloys, and a type of locally made iron hoe.[12] These were not the only forms of money in circulation, although they were some of the most prominent ones used along the primary trade arteries that served overseas commerce.

In their written accounts, Europeans usually overlooked the many types of local and regional currencies that also played a part in maintaining and expanding these commercial networks. Trade was an interlocking system of markets, currencies, and transactions of many kinds, not just those that directly involved the imports and exports on the coast. Caravan travel to and from the interior consisted of transactions over weeks and months, transactions that served to meet the daily needs of porters and merchants, keeping them fed and supplied from this resting stop to that.[13] Commercial traffic depended on the regional, interregional, and more local currencies that operated in the markets and villages far inland, currencies that would be accepted by the many individual vendors that merchants and their entourages met along their way. As we shall see, among the commodity currencies that circulated in all of these networks there were many forms of iron objects, or currency units. They were rarely written about as part of a system of currencies, but luckily many individual examples of them were drawn, described, photographed, and collected by missionaries, ethnographers, geographers, and others. By examining and analyzing their forms and tracing general aspects of their circulation patterns we can begin to see and appreciate why supplies of iron goods were so important to merchant groups engaged in trading over long distances.

Up to now, however, it has been the copper currencies that have held center stage in the scholarly literature, especially the ones originating in the Copperbelt. These are the famous "copper crosses" and other ingots which were produced at least by the thirteenth century. In the nineteenth century these currencies circulated beyond Shaba along trade routes into communities all across the central African savannas. Since this copper was cast into ingots, it had the look of money—recognizably independent units with standardized forms, though sizes and weights changed considerably over time. Ingots like these appear more easily identifiable as currencies and currencies only, for they did not double as commodities. They would have had to be reshaped or remelted to be made into useful products. It is understandable, then, that the cast copper ingots of the Copperbelt would have received by far the most attention of the metal currencies of precolonial central Africa.[14]

Less well known in the literature is the brass wire currency introduced by Europeans into the river trading system in the 1870s. It displaced a prior copper currency that had circulated in the lower and middle Zaire, and was forced onto populations throughout the Zaire basin by the Congo Free State administration.[15] One effect of this new currency was a change in the way wealth was measured and stored. Prices began to be quoted in terms of the *mitako*, the brass wire unit, while ornamental armlets and leglets cast from these brass units began to replace earlier ones that had been made of local iron or copper. These changes followed a certain logic. When a less useful metal was available to serve in some of the more strictly monetary or luxury contexts, such as storing wealth, iron was replaced. But it was never completely replaced, and continued to operate as money in a host of different forms.

Blacksmiths worked iron into particular forms of currency units that were not arbitrary; they were forms that represented deliberate stages in the manufacturing process, from raw metal to semifinished product. Although smelters made metal out of ore, contributing to the total supply of iron money, it was blacksmiths in particular who were in a position to generate and moderate money supply along with its flows and velocity. They were the ones who continually made, remade, reworked, and reshaped iron, from one form into another, and from currency unit into commodity and back. Blacksmiths far outnumbered smelters during the nineteenth century, and most depended on outside suppliers for their iron. They could purchase raw iron directly from smelting sites in their immediate vicinity, or they could rely on merchants to bring them iron that was partially worked by other blacksmiths from farther afield.[16] Especially in the case of iron currencies that were already in semifinished forms, customers themselves might take a unit of money to the smithy to be taken out of currency circulation and made into some desired product. Any blacksmith, by mounting them properly and sharpening their tips or cutting edges, could quickly turn blade currencies into hoes, hatchets, knives, or spears that were ready to use. As the following survey will make plain, blacksmiths, when they set up shop, created not only places of work but also centers of currency production and conversion. Thus, as handlers and managers of iron, a major type of money throughout the region, blacksmiths were engaged in indispensable work.

BAR IRON CURRENCIES

Iron metal direct from the furnace was not yet ready to be worked. Turned from ore into blooms by smelting, iron emerged raw from furnaces in unpredictably various shapes and conditions, sometimes as coherent masses, other times not. The metal itself might be found in small amounts like pebbles dispersed within large chunks of slag, or more conveniently cohered into larger, more easily identified blocks. It was the master smelter who claimed ownership of the prized iron product, and along with it the prerogative to keep his own portion and divide and distribute the rest among his assistants. Portions of blooms could then either be refined and prepared for manufacture into products or sold in their raw state.

In some cases surplus blooms became an article of trade. Accounts of the river networks in late nineteenth-century west central Africa sometimes mention entire blooms for sale that apparently came direct from the furnace, the products of smelters who did not engage at all in smithing but instead supplied other workshops.[17] A missionary traveling on the Aruwimi River purchased one of these blooms, an unrefined conglomeration of iron metal, slag, and charcoal weighing about fifteen kg.[18] On the upper Busira River in 1892, a colonial official mentioned a bloom of iron he saw in a village, though it is not clear whether it had been traded or produced there. It still had some slag in it, and was divided into three pieces, ready to be worked.[19]

More often than not, however, smelters also practiced basic blacksmithing. Soon after a smelt had taken place, they would refine the bloom, hammering it to remove the excess slag and charcoal, and consolidating the bits of iron that had been separated out. Finery operations were therefore often integrated into the smelting workshop as the final stage of work before storing or selling the iron product. And this work required its own special equipment. A stone anvil was considered essential for refining blooms and for the rough work of consolidating or dividing masses of iron, and a well-chosen one was highly treasured. A stone hammer, or in some cases a large iron sledge hammer, completed the primary tool kit of the finery forge.

Smiths who engaged in finery work produced semifinished iron in abstract, three-dimensional forms that are called bar iron. At first glance they resemble ingots, and in some cases they are mistakenly referred to by that term, but it is a misnomer because the forms are not cast. Some workshops that specialized in producing bar iron were described during the dry season of 1874 in the village of Manyara, located in the area between the Lualaba River and Lake Tanganyika known as Maniema. Smelting teams were in operation, feeding ore and charcoal into shaft-type furnaces with up to a dozen pair of bellows working at a time. Two or three of these furnaces were active in Manyara itself, and several in each of the neighboring villages. The ironworkers also refined the blooms and shaped the iron into basic, standardized units:

> After smelting, the iron is worked by smiths into small pieces weighing about two pounds and shaped like two cones joined together at the base. A piece or rod the size of a large knitting needle projects from both ends. In this form the metal is hawked for sale.
>
> Small open sheds are used as smithies, and the anvils and larger hammers are made of stone, but small hammers are of iron. Those of stone are provided with two loops of rope to serve as handles, while the iron hammers are simply grasped in the hand and are without handles.[20]

This description corresponds very well with a particular type of bar iron unit that was identified as a currency and collected as such in the late nineteenth century. Resembling a joined pair of stemmed cones, it circulated along the middle Lualaba and inland from shores on either side.[21] And although its form might seem rather odd and unwieldy, an understanding of conventional sequence in ironworking processes shows otherwise. To a blacksmith the form was not at all abstract. It was immediately recognizable as a convenient, ready-to-work morsel of iron that could be quickly turned into two finished products of certain sizes. The haft (or socket) of a spear, arrow, or knife blade, that is, the pointed stem or end that is meant to be inserted into a shaft or handle, is the first feature that blacksmiths articulate when they make an iron blade. Making the haft first provides them with a way of placing and holding the mass of iron securely on the anvil, for they can grip the haft with pincers or tongs while they draw out and flatten the iron into the proper finished shape.[22] In the case of this particular bar iron, where there were two stems, the central mass would have been partially flattened and then cut in half, with each of the two pieces finally shaped, polished, and sharpened.

This example is only one of a number of bar iron currencies that were made by smiths as recognizably ready-to-work iron, all of them being masses with protruding stems. They were a currency form that was concentrated mainly in the trading networks of the eastern Zaire basin. Although the precise structures of the iron masses differed from one geographical area to another, and could vary in scale within each area, the general principle was consistent. Several other examples that I have identified in museum collections can be cited. In the north, iron masses formed into single stemmed cones and stemmed bulbs circulated along the Uele and Bomokandi Rivers, perhaps as far west as the Ubangi.[23] Blacksmiths along the lower Aruwimi hammered iron into stemmed pyramidal-like masses of many sizes, some of which were then inserted into the hilt ends of fancy knives.[24] Further south, between the Lualaba River and the upper Lukenie, stemmed cylinders served as a form of money and added both weight and ornamentation to luxury knives made there.[25] Single-stemmed bar iron units were observed being made along with finished products in a Maniema smithy in 1876, and were also seen in that same year available for purchase in the thriving entrepôt of Nyangwe on the Lualaba.[26]

Such bar currency units represented a clear stage in the ongoing process of turning iron blooms into iron commodities. And as an intermediate form of iron, refined and stemmed with the precise shape of the blade merely suggested, it made a great deal of sense as a type of money. On one hand, it was a convenient way of storing iron metal for future manufacture, and on the other, it was fungible because it was recognizable as an amount of iron that could be finished into a number of possible products in practically any smithy. In other words, it was visibly both abstract money and potential commodity at the same time, and could operate in all sorts of transactions from monetized ones to barter, moving from currency to commodity networks and even back again if necessary.

Blacksmiths in other parts of the region made a different kind of bar iron, one whose more basic forms did not carry any indication yet as to what kind of product it might be used for. In the west, around the middle Zaire, the central basin, and the lower Kasai, several variants of this more generic bar iron circulated. Although the forms appear to be simple ones, the labor involved in making them could be skilled and impressive. The most stunning example I have encountered so far in museum collections is a perfectly formed iron sphere, a type of currency used in the upper Tshuapa River area. It could have been turned into almost any kind of product, being limited only by the amount of iron contained in its mass (which was not sufficient in this case to produce a sword or an axe).[27] Elsewhere in the river network, in the upper Ngiri, iron was hammered into spherical forms called *ebuni*, which were used as money in assortments of other iron objects. Similar lexical terms for money, though the actual objects are not described, exist in the Ndasa and Tio languages.[28] And Kuba oral traditions mention the tribute assessment of an iron mining area as having been calculated in standard-sized iron balls.[29] Combined, these examples of bar iron in spherical form suggest it may have been a currency that circulated widely along the rivers in the deeper past.

Yet another variant of generic bar iron existed in the Kwango-Kwilu area south of the lower Kasai. Blacksmiths there hammered blooms into flat, rectangular plates that served as money in bridewealth payments, the purchase of property, and probably other types of transactions. Known as *kimbwil* in Hungana, *kimbudi* or *kimburi* in other neighboring languages, much of this bar iron in the late nineteenth century apparently came out of Hungana refining forges, presumably their own smelted product. Mbala smiths claimed also to have made *kimburi*, but it is not clear whether they had actually smelted the iron as well. These iron plates were standardized in form and by name, but not precisely in size; examples in museum collections show them to have had similar proportions but variable dimensions. Notwithstanding these differences in scale, *kimburi* units reportedly passed through and between Pindi, Yaka, southern Mbala, Mbuun, and Hungana communities, and perhaps others.[30] Blacksmiths who produced this currency, like the others mentioned above, thus served extensive geographical areas and multiethnic populations.

Not all bar iron was made by smelters. Some blacksmiths created yet another pattern of ironworking specialization and division of the labor process when they chose to purchase and refine the blooms of other ironworkers. And when they did so, their production of currencies sometimes involved smiths from different ethnic groups. One such example was a smithy observed in early 1906 near the town of Mukoko in Kasai. There, southern Kete smiths were working with iron blooms they had purchased from Lunda smelters on the other side of the Lulua River. These blooms were unrefined, straight from the furnace. Brought back to the Kete smithy, the rough blocks of iron were pounded and hammered with stone refining tools, the slag and other refuse gradually separated out. Switching to another set of tools, iron hammers and anvils, the blacksmiths then shaped the consolidated iron into

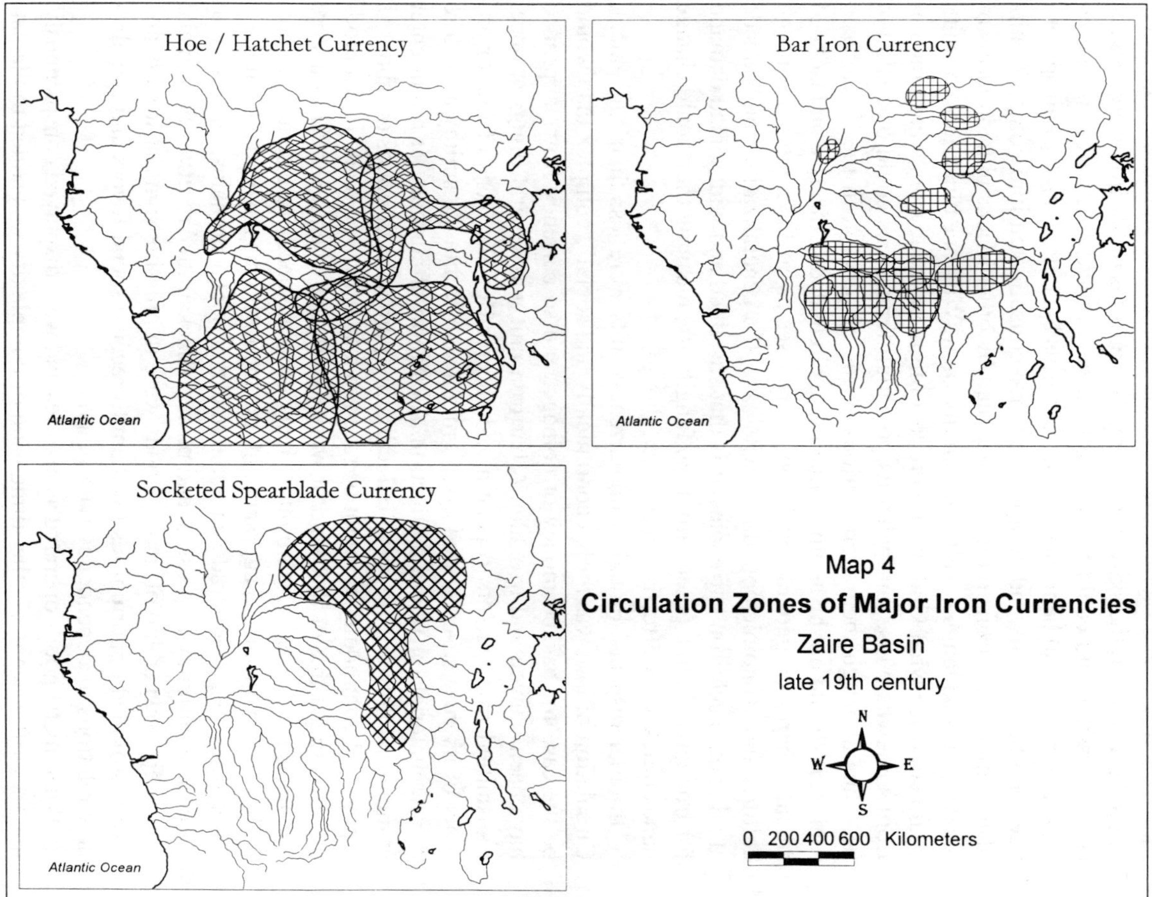

Map 4
Circulation Zones of Major Iron Currencies
Zaire Basin
late 19th century

standard, four-sided lengths called *mburru* [sic].[31] These bar iron units could then be made into different products, or passed into circulation.

It is not possible to calculate how many different smithies were involved in making these various types of bar iron, but there probably were many (see Map 4). Unlike the esoteric knowledge and experience gained through training to be a smelter, the skills for making some forms of bar iron were reasonably accessible and could be learned relatively quickly. Any iron, including recycled iron, could be turned into these forms by a blacksmith or an apprentice with basic skills and equipment. Once consolidated into compact units of iron metal, bar iron could be easily read by other blacksmiths in terms of mass, the visual language all blacksmiths knew. With one quick look, a smith could envisage what kind of iron product might be made from a bar iron unit from somewhere else, and what its potential dimensions could or could not be. Thus although there were various forms of bar iron in the Zaire basin region, it should not be assumed that these circulated in discrete and isolated social spheres or geographical zones. Iron seems to have passed through cultural and economic boundaries with ease. If a generalization can be made at all about such iron currencies, it is that sustained and widespread demand for iron gave them a potentially high fungibility, meaning they could change hands quickly, as effective currencies should.

In other respects, though, iron currency units were less than convenient. Circulation of iron was more economically and socially complex than would be the case for backed money or coinage, and so currency flows appear to have been uneven and irregular. A bar iron unit was several things at once, transmitted from person to person and accepted on the basis of one or another of its guises. As refined iron, it might stay within the smithy as a store of raw material for future production. As a form of currency, it might enter commercial networks to serve in facilitating purchases of goods or labor. As a potential commodity, it might be taken to a smithy to be made into a product used to get this or that kind of work done. And when it served as a store and measure of wealth, it might be lent out to the needy or else used to embellish luxury products, aggrandizing the men who displayed them. In these latter roles, bar iron straddled economic and social domains, operating at times as a currency of prestige and power. Blacksmiths' reserves of iron and elite men's ownership of luxury goods were impressive signs of wealth, and likely provided opportunities for them to receive social favors and establish new relations of dependency with others.

When metal goods operated as luxuries and were displayed conspicuously, and they often were, another dimension of complexity was added to the patterns and flows of currency circulation. This becomes evident if we examine the fancy personal weapon, a class of locally made luxury product worn, displayed, and brandished by elite men all over the Zaire basin during the nineteenth century (see Chapter 5). The hilts, blades, and sheaths of these metal goods were often embellished with forms of currency, especially metal ones. Prime examples of these fancy weapons are the knives that were produced by Tetela smiths and perhaps others around the upper Lukenie and middle Lualaba rivers. I have identified several types of currencies that adorned the hilts of these knives, including at least two types of bar iron units. One was a small cylindrical baton of iron called *mokenga.* Another was the long, rectangular unit discussed above, called *mburru*, which, as an adornment, was folded into a loop and inserted into the end of the hilt. In each case, the bar iron units may well have originated in the smithies of other ethnic groups to the west, along the Lukenie, or to the southwest, in Kasai.[32] Once these currency units were affixed to the knives' handles, they were taken temporarily out of economic circulation and operated instead as important social instruments of prestige. The men who could afford to carry these small, splendid weapons were, in the late nineteenth century, probably clients and allies of Swahili traders, and with such elegant equipage their superior status and refinement would have been visible and apparent to all.

Currencies were used in yet another way: to manufacture the most famous example of fancy iron weapon from this period, the so-called Zappo Zap hatchets. Blacksmiths in Kasai, near Luluabourg, produced extravagant hatchets for the luxury market, and then increasingly for sale to foreigners who could pay premium prices for them. It would be a mistake, however, to see them as brand new inventions that were intended solely for European and North American buyers. To be sure, they did become prominent showpieces in museum collections of central African metalwork, and the makers of them also became well known.[33] But these hatchets were part of a much older, strictly local tradition, their form originating in a smaller, more elegantly crafted prototype that previously had been reserved for prominent Songye political leaders.[34] The ones produced for the luxury market were made less carefully and over time grew much larger in scale and more excessively ornate than the original prototype. Sources agree that these grandiose hatchets were made and used for ceremonial purposes, and designated high rank and status, but there are conflicting accounts about their economic roles. Missionaries and others who lived in the Zaire basin at the turn of the century sometimes referred to these hatchets as "currencies," while researchers have disputed this claim as unbelievable, given the unwieldy forms and heavy weights of the objects.[35]

The dispute can be resolved by looking more closely at the sources of iron used in their manufacture. It was not that these famous hatchets themselves performed as currencies in economic transactions, but rather they were made of money, that is, they were assemblages of bar iron currencies. That this was so can be demonstrated by analyzing the way these hatchets were made.[36] The blades were not the solid, consolidated masses of metal one usually associates with hatchets, but instead were complex openwork structures made out of individual lengths of iron that were looped, twisted, and welded together at the ends. By measuring each of the separate iron components, we find that they exhibit the same forms and dimensions of the long, rectangular bar iron units already discussed above, that is, they are standard iron currency units. Moreover, they remain identifiable as individual units and so could bc easily counted and assessed by a viewer. These hatchets, which clearly were not built to be wielded as either tool or weapon, were instead made to serve as literal representations of wealth.

What this example of luxury production also demonstrates is the degree of specialization in ironworking, and how complex the labor process and social relations could be in the making of specific types of metal objects. Certain iron products were the end result of work carried out by different smiths in different locales and from different ethnic groups, but this organization was created and coordinated by the ironworkers, not merchant capital. An elaborate so-called Zappo Zap hatchet could have been the product of two or three or more separate workshops. Iron was smelted in one place; the bloom was purchased, then refined and turned into bar iron in another; and then certain blacksmiths might purchase a number of these bar iron units, combining, twisting, looping, and welding them together to form hatchets that were signs of wealth and symbols of social status.

Also evident here are the many meanings and values of iron and how they conditioned the ways it traveled from one person to the next. Certainly it passed through cultural barriers, out of one language group into others. It also moved back and forth between money and commodity operations, serving as currency in closing transactions and as a useful raw material in manufacturing. And lastly, the economic values of iron were augmented by social ones. Iron moved vertically, up the social ladder into select circles, when it became a means for men to present themselves as rich, influential, and powerful.

HOE AND HATCHET CURRENCIES

Bar iron circulated alongside other types of metal currency units that were more clearly defined as specific, useful products. Among these were semifinished blades that served in transactions and as currencies of account. Blade currencies could be made from bar iron, or directly from blooms as an extension of the finery process as was observed along the middle Aruwimi river in 1899:

> These ingots [blooms] are 'puddled' [sic] or hammered into bars about one inch wide and half an inch thick, and these are again worked up into blanks for knives, spears, axes, hoes, and such-like things. These blanks or 'forms' have a recognized value as currency, and are distributed over a

very wide area, to be eventually transformed into finished articles, according to the skill and taste of the village blacksmiths and the fashions which obtain in different parts of the country.[37]

Semifinished hoe blades were commodity currency units that found widespread acceptance in the agriculturally-based economies of the Zaire basin. Nevertheless, some scholars designate hoe currencies as "special use" currencies. This designation is based mainly on twentieth-century reports documenting how hoe blades and other iron goods were often used in bridewealth exchanges. Thus, it is assumed, the trade in hoes was always markedly restricted to certain social networks. But the nineteenth- and early twentieth-century evidence for west central Africa shows hoe currencies having circulated more generally and widely, along overtly commercial avenues in addition to the ones for bridewealth. Hoes were noted as a currency in the Uele basin and upper Nile, for example, and were the form of payment assessed by chiefs on traders passing through their lands.[38] Although they may have had restricted circulation in certain places and at certain times, hoes were produced and acquired during the nineteenth century within the extensive market areas of long distance traders. With the imposition of colonial coin and paper money in the twentieth century, hoes and other iron goods continued to circulate (though less vigorously) in the informal economy.

Like other precolonial iron currencies, hoes were valuable both as iron and as a useful commodity. But whereas the generic forms of bar iron currency emphasized the material values of iron, blade currency forms emphasized more specific use values in all kinds of productive work. And these morphological differences affected how and where each type of currency circulated. Bar iron appears to have operated more as a source of iron and a store of wealth. Hoes and other blade forms circulated as potential tools, especially in areas where those tools were in greatest demand. They were practical, since a currency blade became useful with only minimal labor—it needed only to be mounted into a handle and then sharpened for manual use.

Hoes were money for making purchases, preferred components of bridewealth payments, and tools that improved the cultivation of crops. In tropical agricultural production, hoes were important primarily as cutting instruments for clearing fields of small trees, bushes, and vines and preparing the ground for planting.[39] It is these specific tasks one sees embodied in words for both hoe and hatchet, words which tended to be derived from the same verb stem with the meaning "to cut."[40] This set them apart from wooden hoes, which could be used to mound soil for planting manioc, say, and digging sticks, which were serviceable for planting. It was mainly the tasks involved in the preparation of fields that required a sharp-edged tool if the work was to go more quickly. Historical changes in the farming techniques of precolonial Africa are still poorly understood, but it is reasonable to suggest that iron hoe blades, in their role as cutting tools, helped to intensify agricultural production by easing the burden of land clearance.[41]

Hence it is no coincidence that in those areas where there was intensification of agriculture and an increase in the size of investments in women and slaves, there was also a hoe currency in circulation. As wealth from long-distance trading grew over the nineteenth century, there came with it greater densities of settlement, and more focused demand for agricultural produce. Demand was met in part by increases in the scale of agricultural production. "Big men" who got their start in trade could grow bigger by investing in slaves and women, thereby turning their households into firms that supported their trading operations. It was the production of manioc, especially, that literally fed the caravan traffic, plantations of it located at intervals along the long-distance routes.[42] And iron fed the wealthy merchants, who needed hoes to use in their trading expeditions, for acquiring women, and to equip their farm slaves.[43]

The economic effects of hoe currency circulation were similar to those of bar iron, in that they facilitated transactions linking regional commerce with local economies. For members of a caravan, carrying along an investment of hoes offered some security, since they were durable goods that could purchase foodstuffs and other necessities. For rural smelters and refiners, manufacturing currencies made it possible for them to tap into the wealth that was growing along the axes of external trade. Evidence that a certain amount of economic integration was taking place is indirect but consistent and persuasive. It shows up in the varieties of tools used by farmworkers and the names for them shared among different languages. When these data are plotted on a map, one sees broad areas of trade in hoes that extend well beyond the major routes of external commerce (Map 4).

At least four such hoe currency zones covered almost the entire Zaire basin. One of them spanned portions of the eastern forests—within the trading spheres of Swahili-based merchants and their clients. Several types of flat blade circulated as currency in this area, all called *shoka*, a term meaning "hatchet" in Swahili (Fig. 4-1). The distribution of the lexical term follows the Lomami River trade route, and angles across Maniema into the Lakes area. The Swahili term for hoe is *jembe*, but it has rarely been used or accepted into languages of the Zaire basin.[44] The types of blades that came under the rubric *shoka* were variable, and were described at times as either hoe, hatchet, or spear blades. It is not clear whether these attributions, mostly by Europeans, are faulty, or if the term *shoka* came to be used as a general one for imported blades or trade iron. Most examples were designed for cutting or chopping,[45] though several were pointed spear or hoe blades.[46] They operated as a currency especially for purchasing foodstuffs, again underlining their importance in long distance trade:

> All the smiths in the Stanley Falls district depend on these [local shokas] in making axes, knives, spears, arrowheads, etc., and households depend on them as a market currency. All our Yakusu food for work-people and children is paid for with shokas—three bundles of plantain = one shoka—and without these we are in difficulties at once. When I was at Yakusu early in the year the stock was running low, but we thought the interruption of the supply was only temporary, and that the Arabs [sic] and others would soon be coming along as before to exchange their shokas for cotton goods, enamel ware, and such-like things which we import for exchange purposes. As it turns out, however, the new railway to the south is absorbing all the supply, and the government and traders are being compelled to import from Europe. So far, the most popular substitute for the native-made article is one cut from a rolled plate of iron and not worked at all. Some have been sent that are slightly thicker in the middle in imitation of the native pattern submitted, but has not caught on at all, some say because of the rib in the middle. I think the fault is not with the pattern, but with the quality of the iron, the natives requiring an article that will not run to slag [sic] into their charcoal pits as it is being worked up.[47]

Some of the blades designated as *shoka* were too light in weight to have been used for agricultural tasks; several of these would have had to be consolidated to make a heavier tool. Others were flat sheets of iron that could have been welded together to make clapperless bells, important signaling and message devices. In any case, the units had standard dimensions and could be used in multiple ways, even as a source of iron for making other products, as the description above indicates.

Another hoe currency zone grew around trade and tribute networks in the eastern savannas. Throughout this region, hoe blades circulated over an area almost as large as the circulation zone for the famous copper cross ingots. Patterns of production and circulation are especially difficult to trace here, for no single, specialized merchant group emerged in Shaba to monopolize commerce and expand currency zones. However, many of the tribute collecting and transit trade networks involved speakers of Luba and Lunda languages, and lexical evidence indicates that hoes were among the goods exchanged along these networks.[48] At least a dozen contiguous language groups share a term for hoe, *lukasu*, that may well have originated in Luba-Shaba or nuclear Lunda.[49] So far I have not found evidence of a prominent center of production, but there are scattered references to hoe manufacture in several locations west of the Lualaba. Smelters working in Shaba often estimated iron output in numbers of hoes, while Ikalebwe (Songye) smelters shouted for hoes and axes to come out of the ore while their furnaces were in operation. Luba traders carried hoes eastward, according to one transit trade pattern, where Hemba dealers accepted them for salt.[50]

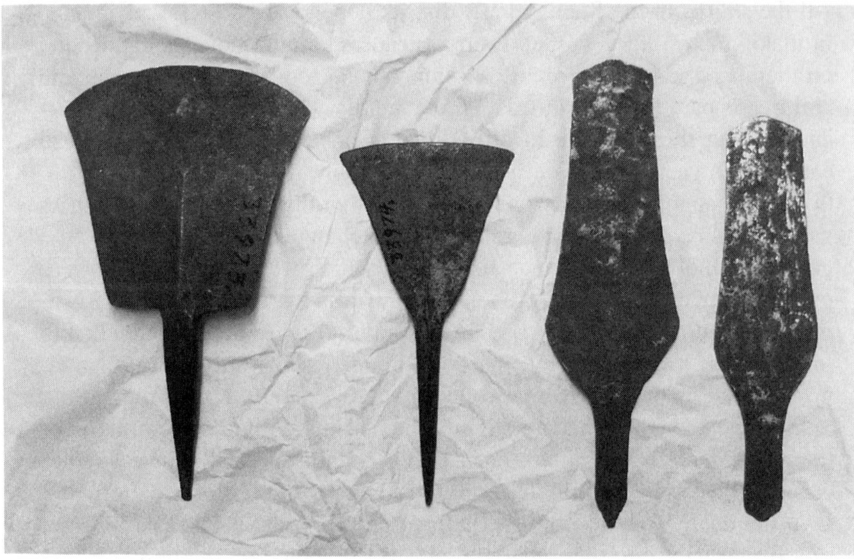

Fig. 4-1: Iron hoe (*lokongo* or *longo*) and hatchet (*shoka*) currency, Zaire basin, late nineteenth century. On the left, a hoe design new to the Lake Tumba area; second from left, the older hoe design to that area; both made by Sakanyi and Ntomba smiths. On the right, two examples of currency used in the area around Stanley Falls, upper Zaire. 24 cm and 21 cm long. The Field Museum, Chicago, catalogue numbers (left to right), 33973, 33974, 33975, 33976. Collected by Joseph Clark between 1894 and 1903. [Photograph by the author]

A third hoe currency zone extended from Bobangi networks in the Zaire River trade. Population densities were relatively lower in the rainforest areas, and intensified food production was probably not as widespread as it was in the savannas. Hence the iron currency forms there included also arrow, knife and spear blades, a variety reflecting different patterns of demand.[51] Agricultural techniques tended to rely more on using multipurpose knives for clearing fields and other tasks, and it was in rainforests where cultivation with digging sticks continued into the twentieth century. Nevertheless, there is evidence that traders plying the river networks circulated hoes along with other types of iron blades relatively recently, though the patterns are difficult to discern because the languages spoken there are so closely related.

Corroborating evidence comes from certain reliable collectors of the blades themselves. In Ntomba country along the middle Zaire, for example, at least three very different types of hoe blade coexisted in the late nineteenth century. One was the hoe blade called *lokongo*, its form very much like ones I witnessed recently being used by women farmers in nearby Ekonda communities.[52] It was described as a "new" form of hoe in the 1890s, made by Sakanyi and Ntomba smiths. Another one, smaller, triangular in form and curved in profile, was referred to as an older type of hoe also used in the area (Fig. 4-1).[53] The third one had a completely different design, oval in silhouette, with a slight midrib in the center of the blade. This same type of hoe was an iron currency along the Uele river, over 600 km away.[54]

Finally, an important hoe currency zone involved Ovimbundu merchants and the trade routes they traversed between Angola and their suppliers inland as far as the Lunda kingdoms. Within this zone, the most common type of hoe that circulated was oval in form, angled rather than flat across, and called *temo*.[55] One prominent center for making them was Ndulo, located in the central highlands of Angola, not far from some of the main overland caravan routes.[56] In and around Ndulo, ironworkers organized many individual workshops for high volume hoe production, combining several strategies for intensifying their output. First, family and probably slave labor was recruited in the mining, transporting, and preparation of ore for smelting. Second, production was year-round: charcoal making, mining, and smelting of the iron ore were done during the dry season from May to September; hammering out the hoes took place during the wet season. Third, certain tools were designed specifically to speed up the manufacturing process. Blacksmiths worked on a stone selected for its versatile surface qualities, one that had both flat areas for drawing out the iron, and special depressions that served as molds for forming the curvatures of the hoe blade. Their sledge hammer was also versatile. Its rounded side served to pound the iron flat in the drawing out stage, while hammering with its pointed end aimed downward helped to form the angle along the center axis of the hoe.[57] This work required strength and precision, and had to be resourcefully choreographed. Blacksmiths saved energy by using specially chosen and designed tools, which if used correctly reduced the production process to a minimum number of hammer blows.

The people living in Ndulo and surrounding communities were well aware of what the making of hoes meant in terms of their well-being. They rendered homage to the sledge hammer by addressing it with the title *osoma*, the designation for a chief, explaining that both were providers of food.[58] They meant this in the larger sense, for not only did the hoes serve directly in agricultural production as a tool, but volume production of them brought in other forms of wealth. Hoe manufacturers were linked to both local and regionwide markets, for they could sell their goods for food and other produce, or to merchants and others for different forms of currency. A missionary, writing from Chisamba in 1894, described how he built up a flock of chickens with hoes:

> Last dry season we could scarcely get a chicken for an invalid. I therefore bought up all the old hoes and turned them into our blacksmith's shop to

> make into irons for the cart, etc. I then sent to Ondulo [sic] and bought a lot of new hoes, each one worth a good-sized chicken. Now the rainy season has begun the women must have hoes to work their fields. I was therefore able a few days ago to place a lot of fowl in my yard and still the people come wanting more hoes. We now have quite a supply [of chickens] for our table at a very small cost.[59]

Thus it can be inferred that Ndulo was a center of currency production and also currency conversion. This missionary's initial purchase of hoes was probably made with the European manufactures that served as currency in long distance trade, such as imported cotton cloth. Then farther away, the offer of a hoe was a way to convince a rural or village farmworker to part with food stores, say, a chicken. Monetary theorists, who tend to avoid confronting the complexities of monetary experience, would consider commodity currencies inefficient and a hindrance to the integration of urban and rural economies. But in tracing these transactions it becomes apparent that some economic integration was indeed taking place, between what might otherwise appear to be separate external and internal currency spheres. When ironworkers produced multipurpose iron goods they were making such currency conversion and economic integration possible.

Ovimbundu merchants and others who bought, carried, and spent hoe blades left indelible evidence of their transactions in the toolkits used and languages spoken within their trading spheres. Although the precise types of hoe blade differed according to the workshops that produced them, merchants called them all by the name most familiar to them in their own language. When speakers of other languages purchased new types of hoes from these merchants, sometimes they imported new words for them as well.[60] A prime example is the Cokwe language, which includes several different lexical terms for hoe. Two of them, defined as "indigenous," were apparently well established, familiar tools. Two others were designated as imports, one from Ovimbundu and another more recent one from overseas.[61] These different words are the residue from a time when Cokwe patriarchs were important producers and consumers of iron. Some were successful ironworkers, some were suppliers of wax, ivory, and rubber to caravans in the second half of the nineteenth century. Most were also wealthy investors in slaves and women. Hoes were not new in the nineteenth century, but there were more of them made and circulated then as currencies over longer distances.

Tool designs, types of iron blades and different kinds of handles, followed certain cultural preferences, but these preferences were subject to change. The Ntomba and Cokwe examples already cited, where new blades were incorporated through trade, were not isolated ones. It was common for several types of hoe to coexist in the same locale, a phenomenon which reflects complexities of local demand—how new social groups were created through large slave holdings, for example, and how changes in farmwork were instituted as agricultural production intensified. A brief description of agricultural work along the Kwango in 1893 specified two very different types of hoe. One was an iron blade mounted like an axe on straight handle; the other was a different type of blade, mounted on a forked, double handle, a feature associated with hoes used in Mbundu society.[62] Cultural preferences were strong enough to register as such, products having been associated with this or that social group, but they were not immutable. With greater capitalization of agriculture and slavery came social and cultural transformations that expanded the zones in which blade currencies flowed.

KNIFE AND SPEAR CURRENCIES

Iron blades were made in so many forms and traded and modified so frequently that it is sometimes misleading to refer to them specifically according to how they would be used. Broad blades with wide cutting edges could be mounted one way and serve as a hoe, mounted another way to serve as a hatchet. Central African tools that show up in museum collections display the multiple potential uses of blades quite clearly to the careful viewer, as do vocabulary items such as the so-called *shoka* currencies discussed above. Knife and spear blades provide yet another example of the ways blades could be adapted for different uses simply by mounting them differently, on either short handles or long shafts. And by tracing the locations where such blades were made, described, or collected, no matter how they were fitted for use, it is possible to show how widely some of these types of blade currencies circulated.

Knife and spear blades were currencies mainly in the northern and eastern rainforests of the Zaire basin, though they were also important articles of trade in the central basin. Both were used in various capacities as personal weapons and as tools in food production, the knife in agriculture and the spear in hunting. And they were usually designed along similar lines as cut-and-thrust implements, pointed at the tip with cutting edges on both sides. Knife blades were always hafted at the end for insertion into a wooden handle; spearblades, though, could be equipped with either a sharp haft or curved socket. Hence they were often, though not always, interchangeable.[63]

Various types of blades originating in the north were traded as currencies southward and into the main river system. In the rainforests, the general trend was for iron currencies and other products from the Ubangi and Uele River basins to be exported into the upper Zaire, and from there southward along one of two main routes. On the west side, river merchants carried goods down the Zaire as far as Malebo Pool, while on the east side, river traffic headed up the Lomami and Lualaba Rivers. How old this trading pattern was remains to be established, but certainly as the nineteenth century progressed, commerce along these routes increased in volume and frequency.

One type of semifinished blade that served as money in the Uele basin was rather heavy and substantial, both in the amount of iron it contained and in the degree to which it was a recognizable commodity. As either a knife blade or a spear blade it was unusual in form. Not having the pointed tip common to cut-and-thrust blades, it instead broadened gradually into a flat or gently curved spatula-like termination. As far as I know, they were always hafted, never socketed. Such blades have been referred to in the published ethnographic literature as "knife money" or spearblades, and they were currency for bridewealth payments, purchase of foodstuffs, and other transactions in the Azande and Mangbetu areas of the Uele basin.[64] They were included as part of a palaver fee displayed in the village of Manziga, photographed early in this century (Fig. 4-2).[65]

Merchants traded these currency units southward along the western route toward the middle Zaire, where they became the blades for knives of prominent and wealthy men. Such fancy knives will be discussed more fully in the next chapter, but here they serve to illustrate well the problems of categorizing some of the blacksmith's products by use. Names for the currencies that were recorded by collectors refer to them as "hatchets," *mangua* in Zande and *nékombi* in Kere.[66] This terminology stresses the cutting function and explains the broad, flat-edged tip. However, the blade was too long to be mounted like a hatchet, and its haft was probably inserted onto a long shaft, as a hatchet-spear. As such, it would have been particularly lethal as a weapon, given its mass and long cutting edge at the tip, and it probably served at one time for hunting large game, including elephant.[67] Its haft could also be mounted into a short handle in order for the blade to be used as a knife or sword, with its lengthwise edges sharpened. Many such examples existed in the middle Zaire and central basin by the middle of the nineteenth century. This case illustrates very well how blades could be defined and redefined, the very same type as a hatchet-spear in one place, and as a knife-sword in another.

Most of the other northern blade currencies flowed southward along the eastern route. One of them, which was used near and around the Uele basin, resembled a flattened unit of stemmed bar iron. Referred to most often as *mapuka*, or spear money, the object was also called *mbeli*, knife money, perhaps by Ngombe traders and others.[68] It was noted as a currency in Azande, Mangbetu, and Barambo communities, though it was accepted also southward, at least as far as the Zaire River.[69] Since it was semifinished, it was still not yet a fully recognizable commodity and, as the lexical terms suggest, could have been converted later into either a knife or spear blade or into two small blades. A similar type of stemmed blade, perhaps historically related to this one, was used alongside the single stemmed bar iron units much further south in Kusu country between the Lomami and the Lualaba.[70]

Several types of socketed spearblades traveled as currency through the eastern trade networks of the Zaire basin. Out of the entire range of spearblades

Fig. 4-2: Spearblades lined up as a palaver payment, Manziga village, Niangara, Mangbetu kingdoms.Negatives/Transparencies #224066, Courtesy Department of Library Services, American Museum of Natural History. [Photograph by Herbert Lang, American Museum of Natural History Congo Expedition 1909-15]

produced in the region, only certain specific forms were used as money. The most prominent examples were referred to in the ethnographic literature as *konga* or *dikonga*, a general and widely used term for spear. Of all the blade currencies discussed so far, these were the ones closest to being finished products. Not only did they have completed sockets at one end, but they were rather ornate. Often the surfaces were decorated with patterns of grooves or incised lines along the left side of the blade's face. They varied considerably in scale, though it appears that most of the original prototype spears were not particularly large and had been developed by ironworkers in the Uele and Aruwimi basins. Moreover, by the end of the nineteenth century, iron imported from Europe was being used to manufacture certain types of these blades, which raises questions about variability and change in the values of iron, which will be discussed further on.

IRON AS MONEY, SMITHS AS MONEY-CHANGERS

Just as assortment bargaining on the African coasts included commodity items from external and internal trading spheres, assortments in the African interior were made up of a variety of commodities, local and foreign. Among these assortments there were iron commodity currencies, combinations of which varied from place to place and over time. Certain units appear to have been manufactured in various standard sizes and scales that could be counted out in combination with others. For example, where there is evidence enough to make calculations, we find that blacksmiths made differently sized units that worked together, with fixed numbers of smaller denominations being the equivalent of a larger iron unit.

How these different units of iron could work as a money system was well described to a missionary in 1892 by male elders in what is now Gabon:

> This is the old iron money of the Fang. A generation ago all transactions from buying a plantain to buying a wife could be carried out with this money. *Eki* means iron. The regular plural of which is *biki*; but that plural came to have a technical meaning just as coppers have come to mean coins of a certain kind. The Fang smelt iron and make it into all sorts of knives, tools, etc. But such work required a degree of skill possessed by but a few so there came about a division of labor, a principle so little understood and practiced by savages [sic]. Iron was smelted by a few and beaten into any rude shape. This iron was sought by others who turned it into spears, axes, adzes, which everybody wanted. Hence the principal article of trade becomes iron, the one thing which every man could not get for himself. In some way it came to be put upon the market in this form which they all call *biki*. So many of these little irons would make a spear and so many more an axe. Anyone who wished to make one of these articles bought

> enough *biki* and beat it into that form. This fixed the value of these little bundles of iron. But even savages [sic] feel the need of a medium of exchange and the value of tens and hundreds of these *biki* being fixed a person who did not wish to make a knife or spearhead would take the *biki* in exchange or seek it even that he might have on hand some exchangeable commodity . . . *Biki* was to them what money is to us. Today only old men can explain how they were used. Formerly it was carefully counted out; two wrapped together and five of these = ten *biki*. Ten tens = one hundred *biki*. Ten hundreds = 1000 *biki*. 500 *biki* and a number of spears, knives, etc. was the dowry [sic] of a woman.[71]

The Ogowé basin was not the only place where small denomination iron units operated, and it may have been a common development. In the upper Lukenie, it was said that two small *kundja* blades were equal in value to one large one,[72] and when I weighed examples of these objects I found that this equivalence was based on the amount of iron in each object (weight serving as a proxy for mass). In other words, the iron mass of two small blades could make one large one. In the Azande and Mangbetu polities in the Uele basin one finds a similar pattern. The heavy hatchet-spear blades that served as a large currency unit were stated to be the equivalent in value of five *mapuka* units.[73] And indeed, the weights of the large blades range between 500 and 800 g, while five *mapuka* would together weigh between 450 and 650 g.[74] Bar iron in the form of single-stemmed cones circulated alongside these blade currencies, and they weighed half of one *mapuka*, suggesting an integrated system of equivalencies from small units to large.[75]

There is also evidence that small units of iron were not just made for local transactions but were also circulated in regional exchange networks. Libinza traders who worked the Ubangi, Ngiri, and parts of the upper Zaire river basins used them in their commercial expeditions. Y-shaped iron objects known as *kwa* served as small denomination currency units, and a center that produced them was located on the Ngiri River. Ore was imported from the Zaire River networks, and iron metal flowed back into the trading system as money.[76] The units came in two sizes, one weighing between six and ten g, the other about twenty g. Values of heavier iron currencies were quoted in *kwa*, again based on the number of them it would take to make the heavier iron object. One heavy type of knife blade currency was valued at twenty to thirty *kwa*, and examples of it weighed between 220 and 470 g, the approximate weight of twenty to thirty *kwa*. Another type of blade currency, this one heavier (between 480 and 610 g), was quoted as having been worth 50 *kwa*.

I enumerate these details to show not only that there was an iron currency system and how it worked, but also that it was developed and managed by blacksmiths, their manufacturing processes being the keys to understanding it. Iron currencies were made into units based on the mass of metal and how much mass was necessary to make a larger currency unit or final product. From bar iron sources to blade currencies to finished products, from the small currency unit to the larger, iron currency forms were made out of divisions in the manufacturing process. Blacksmiths did the dividing, generating something like assembly line production, but instead of it being consolidated under one roof or controlled by one investor, it was dispersed in space, in separate workshops. Blacksmiths were the pivotal figures in a commodity currency system where iron could move from smithy through social and economic networks and back into the smithy then back out again. Their iron currencies did what currencies are supposed to do—move goods from place to place, close deals, facilitate services, and get work done.

Many people must have profited from this decentralized currency system. Large denominations were used in large transactions, especially those that were more closely tied with external markets and higher priced goods. Smaller denominations were used in assortments for bridewealth, for market purchases, and for buying foodstuffs. The various forms and values of iron currencies thus bridged rural and external economies, integrating entrepôts with peripheries. Even women, who were denied shares of blooms, could gain access to iron by selling their own produce for a unit of bar or blade currency. Iron wealth, like trade wealth, was concentrated in the hands of men, but iron commodity currencies could work to redistribute that wealth to some degree.

Above all, it was the blacksmiths who profited from this currency system by making money and also by changing it. The small denomination units were accountable as independent units, but if they were to be changed by a blacksmith into a larger object, they could no longer be accounted for in the final product. If a man brought thirty *kwa* to a smith to be made into a knife, who was to say in the end that all thirty had been welded together? It is likely, based on what was observed with brass *mitako* units, that blacksmiths took small transaction fees for converting iron currencies.

Another way blacksmiths profited, at least in the short run, was by incorporating iron from overseas into the system. Imported iron (the forms and amounts of it that were traded in the Zaire basin both before and during the nineteenth century) is an important issue, but it is far too complex to be discussed at this time.[77] However, certain examples can shed light on the issues that are the focus of this study, in particular the valuable material properties of iron and blacksmiths' preferred sources for it. Smiths were continually assessing and evaluating their changing supplies of iron, including iron from Europe. They identified one particular type of imported iron as being soft and malleable, and by the 1880s smiths all over the region sought it out. Iron hooping that was used to wrap bales of goods shipped to Africa was available wherever Europeans were—at colonial posts, trading centers, or missionary stations[78]—and because of its malleability it could be consolidated by welding and made into currencies. A blacksmith in San Salvador used it for just this purpose:

Fig. 4-3: Topoke men with large currency blades. Copyright © British Museum, Negative number MM005282. [Photograph by Emil Torday, British Museum Congo Expedition 1907-9]

> Hoop-iron was an article of trade at that time, regularly imported [1879]; Miguel obtained it, folded it into lumps of the proper size, heated it in his little furnace with charcoal—specially prepared for the purpose—and having welded it, beat it out into the large hoes which were then in fashion. At the back of the hoe was a spike which was heated, and so burnt into a wooden handle. A most serviceable implement was the result. There was a great demand for these hoes, especially in the dry season, when the new ground was broken up, and Miguel worked early and late. His hammer might be heard from four o'clock in the morning until ten o'clock at night.[79]

Thus imported iron, along with recycled local iron, provided a way for enterprising blacksmiths who did not have smelting or refining skills to enter currency markets.

Imported bar iron from Europe also provided new opportunities for smiths to manufacture very large-scale currency blades. Long poles of industrially produced iron were cut into smaller lengths by blacksmiths in the eastern side of the basin, then traded as small bars or made into blades similar in scale to the ones made from local iron.[80] But in some cases these long poles were used as a single source of iron, and smiths made them into large lances and spectacular spear blades from 1.5 to 2 meters long (Fig. 4-3). The source of iron for these enormous currencies has never been identified because the method of manufacture was apparently not of interest—but there can be no question that they were made from these imported bars.

Known by either the term *liganda* or *ngbele*, the huge spearblades were associated with Topoke and Kele river traders of the lower Lomami, and began to be noted and collected by Europeans in the 1880s and 1890s.[81] Now they are in many museum collections and are often pictured as "typical" examples of "traditional" iron currency in central Africa. Although the manufacturing process did follow a rather typical pattern of blacksmiths using bar iron from elsewhere, and although their forms were modeled after locally invented prototypes, these currency blades were not at all typical in scale. They do show, however, that European iron could be absorbed into central African economies as simply yet another source of raw material for blacksmiths to work with. And, given the numbers of Europeans who were dazzled enough to want to purchase these unwieldy blades, smiths who made them probably turned a nice profit.

But as the nineteenth century drew to a close, ironworkers' influential social positions as suppliers and managers of money became increasingly paradoxical. On the one hand, the limited volume of bloomery iron production restricted the supply of money and kept iron valuable during a time of economic expansion. On the other hand, blacksmiths turned imported iron into currency by reforging it into recognizable commodity currency forms. Individually and in the short run they gained. But in the long run they unwittingly contributed to the undermining of their own occupation by working to devalue iron. For while they were converting external currencies into internal ones, they were also participating in the playing out of Gresham's Law, where bad money drives out good. Most iron from overseas was relatively cheap to produce. However, it was of lower quality than bloomery iron and it was not well suited to blacksmithing. Even when it was disguised in local forms it gradually debased iron and lowered its value. Central African blacksmiths in the twentieth century eventually joined the blacksmiths of Europe and North America in lamenting the inferior material qualities of cheap, industrially produced iron—made for all sorts of purposes, but not for smithing.

NOTES

[1] Korse, "La Forge," p. 34.

[2] Galbraith observes shrewdly that throughout history, monetary and financial experts, like doctors, have cultivated the belief that much of their expertise comes from extraordinary and occult powers. John Galbraith, *Money. Whence it Came, Where it Went* (Boston: Houghton Mifflin, 1975), p. 4.

[3] This is in contrast to coinage, which usually bears visual evidence of the political regime backing it (e.g., portraits, emblems, mottos, etc.).

[4] *New York Times*, Wednesday June 12, 1996. Values in dollars per pound are: $61.56 for silver; $1.08 for copper; $.08 for pig iron.

[5] Even if smelts produced iron with high carbon content, the iron would lose carbon in the refining and forging processes. Personal communication, J.E. Rehder, 13 March 1989. See also Manfred Eggert, "Katuruka und Kemondo"; and Idem, "On the Alleged Complexity of Early and Recent Iron Smelting in Africa."

[6] Oppi Untracht, *Metal Techniques for Craftsmen* (London: Robert Hale, Ltd., 1969), p. 271.

[7] Kriger, museum notes, technical data from MRAC Acc. # 14531, 14532, 14533, 179 2/1 and 2/2; MRAC Acc. # 14528 8/1 to 8/8, 177 2/1, 45241 6/1 to 6/6, 145240 3/1 to 3/3.

[8] Four *ikonga* blades would weigh together between 560 and 720 g; cast copper crosses weighed between 600 and 1000 g. Kriger, museum notes, technical data from MOM Acc. # 1909.Ty.698; MRAC Acc. # 23754, 23212, 38072, 38073, 73.16.24. The equivalence was taken down by Torday, Congo Expedition, large format MS, p. 873.

[9] Kriger, museum notes, technical data: *kundja* blades weighed about 80 g in the small version, 160 g in the larger version; they are very similar to the blades called *budju*; for the *boloko* units, see MRAC Acc. # 69, 19118, 54.84.10, 19464, 19334, 19116.

[10] Exceptions are the work of Georges Dupré, most notably "The History and Adventures of a Monetary Object of the Kwélé of the Congo: Mezong, Mondjos, and Mondjong," in *Money Matters. Instability, Values and Social Payments in the Modern History of West African Communities*, ed. Jane Guyer (Portsmouth, NH: Heinemann, 1995); and J. Guyer, "Indigenous Currencies and the History of Marriage Payments," *Cahiers d'Etudes Africaines* 104, 26–4 (1986). Earlier surveys by Einzig, Quiggin, and Mahieu did not include most of the examples of iron currency units, but focused mainly on copper and copper alloys among the metals. Paul Einzig, *Primitive Money* (Oxford: Pergamon Press, 1966. Second edition, enlarged and revised.); A. Hingston Quiggin, *A Survey of Primitive Money* (New York: Barnes and Noble, 1970. Reprint of 1949 edition.); and Adolphe Mahieu, "Numismatique du Congo," *Congo. Revue Générale de la Colonie Belge* 1 (1923).

[11] For a theoretical overview of commodity currencies in West Africa, see James L. A. Webb, Jr., "Toward the Comparative Study of Money: A Reconsideration of West African Currencies and Neoclassical Monetary Concepts," *International Journal of African Historical Studies* 15, 3 (1982).

[12] "La Monnaie," *Le Congo Illustré* 9 (1892), pp. 34–35.

[13] The Lwena were important suppliers to Ovimbundu caravans, providing agricultural products, fish, and iron. Vellut, "Notes sur le Lunda," p. 131.

[14] See Bisson, "Copper Currency"; de Maret, "L'Evolution Monétaire"; Schoonheyt, "Les Croisettes du Katanga"; and de Maret, "Histoires de Croisettes."

[15] See R. K. Eggert, "Zur Rolle des Wertmessers (*mitako*) am oberen Zaire, 1877–1908," *Annales Aequatoria* 1 (1980).

[16] In the Aruwimi River basin, some smelting groups relied on merchants to supply them with mineral ore; smiths who did not smelt relied on merchants to supply them with bar iron to be made into other products. One observer stated that commercial relations for industry were as frequent as those for agriculture. Joseph Halkin, *Les Ababua* (Brussels: de Wit, 1911), p. 510; L. Védy, "Les Ababuas," *Bulletin de la Société Royale Belge de Géographie* 28 (1904), p. 204.

[17] Nahan, "Reconnaissance de Banalia," pp. 557–58.

[18] Pictured in Johnston, *Grenfell*, Vol. II, p. 798.

[19] Charles Lemaire, *Congo et Belgique* (Brussels: Bulens, 1894).

[20] Verney Cameron, *Across Africa*, 2 Vols. (London: Daldy, Isbister, and Co., 1877), Vol. I, pp. 371–72.

[21] Kriger, museum notes, technical data for MRAC Acc. # 33703, from the Lusamba area, weighing 530 g.

[22] Kriger, field observations, Lopanzo, 1989; Emma Maquet, *Outils de Forge du Congo, du Rwanda, et du Burundi* (Tervuren: MRAC, 1965), p. 42.

[23] Kriger, museum notes, technical data for the Uele, MRAC Acc. # 19499 15/1 to 15/15; for the Bomokandi, MRAC Acc. # 2415 3/1 to 3/3; and for the Ubangi, MRAC Acc. # 20896 and 20897.

[24] Kriger, museum notes, technical data for bar iron attached to hilts of knives: ROM Acc. # HA 2480; MRAC Acc. # 900 3/2, 900 3/3, 55.47.1, and 56.79.19.

[25] Kriger, museum notes, technical data from examples collected by Torday: MOM Acc. # 1909.Ty.555, .556, .557; 1909.Ty.769, .782, .794. See also MRAC Acc. # 437 4/3, 6477, 23254, 23255, 16114, 16115.

[26] Stanley described them in the first instance as "iron balls with spikes" and in the second instance as "iron knobs." Henry Morton Stanley, *Through the Dark Continent*, 2 Vols. (Toronto: Magurn, 1878), Vol. II, pp. 121, 142.

[27] MRAC Acc. # 29.

[28] Mumbanza mwa Bawele, "Les Forgerons de la Ngiri," p. 132, fn 16. For lexical terms: Ndasa (B25) money = *ebone*; Tio (B75) money = *luni; ibuunu.*

[29] Vansina, *Children of Woot*, p. 141.

[30] Klein, *Frobenius*, Vol. I, pp. 45, 313; de Sousberghe, "Forgerons et Fondeurs de Fer," pp. 27–28. In Yaka communities, this currency was said to have been originally made by the Samba or Hungana. Kriger, museum notes, technical data from examples in collections: MOM Acc. # 1907.5-28.99; 1907.5-28.133; 1907.5-28.120; MRAC Acc. # 43822, 43823, 53.74.3449. The languages referred to here are contiguous, belonging to the B, H, and K Bantu language groups.

[31] Klein, *Frobenius*, Vol. III, pp. 41, 170. This spelling is from Frobenius, and it may be related to the term *kimburi*, though the forms of the bar iron are different.

[32] Many examples of these knives exist in museum collections, usually identified as Tetela, but they could also have been manufactured by Kusu smiths or others. While the term *mokenga* is from the Sakata language, these bar iron units could have been made by other smiths as well. For examples of knives with these particular units attached, see Kriger, museum notes, technical data from MOM Acc. # 1907.5-28.356; MRAC Acc. # 24851, 23259, and 16113. For examples with loops of rectangular bar iron, see ROM Acc. # HA 2463; MOM # 1909.Ty.848; and MRAC Acc. # 16100, 22253, 23848, 22252, 16103, and 8157. Mahieu mentioned that "sticks" of iron were used as money on the Lomami River, but he did not describe them further or include an illustration. Mahieu, "Numismatique," p. 658.

[33] They were referred to by most Europeans as "Zappo Zap." The name Sapo Sapo or Nsapo refers to the name of a chief among the Ben Ekie section of the Songye, who relocated his followers several times in the 1880s, finally settling near the Congo Free State station of Luluabourg in 1887. He died in 1894, and was succeeded by his son, also sometimes referred to as Zappo Zap. See *Biographie Coloniale Belge*, Vol. III, columns 937–39; and P. Timmermans, "Les Sapo Sapo près de Luluabourg," *Africa-Tervuren* 8, 1-2 (1962). For the Ben Ekie, see Nancy Fairley, "Mianda ya Ben'Ekie: A History of the Ben'Ekie" (Ph.D. diss., State University of New York at Stony Brook, 1978).

[34] Torday brought back one example, stating it was the older type of Songye axe, and emphasizing that it should not be confused with the so-called "state axes" made by the Nsapo or Zappo Zaps. See MOM, accession documentation for 1908.Ty.976. Another example is in the MRAC, Acc. # 55.67.2.

[35] Quiggin, *A Survey of Primitive Money*, pp. 63–64.

[36] I have examined over 100 hatchets of this type. Examples are: ROM Acc. # HA 120, HA 1220; MRAC Acc. # 2754 6/2, 8093, 65 16/3, 65 16/1, 6444.

[37] Grenfell's diary, Nov. 1899, published in Hawker, *Life of Grenfell*, p. 446.

[38] Georg Schweinfurth, *Artes Africanae. Illustrations and Descriptions of Productions of the Industrial Arts of Central African Tribes* (Leipzig: Brockhaus, 1875), Part II, fig. 20; Part IV, figs. 14, 15; and Wilhelm Junker, *Travels in Africa during the Years 1875–1886*, 3 Vols. (London: Chapman and Hall, 1890–92. English translation of 1889 German edition.), Vol. III, p. 49.

[39] Marvin Miracle, *Agriculture in the Congo Basin* (Madison: University of Wisconsin Press, 1967), pp. 36, 41–42.

[40] For example, the root form **-témo* (Guthrie CS 1705, 1706), an old term for "hatchet," later becoming a term for "hoe," derived from the verb stem **-tém-*, "to cut." See Vansina, *Paths*, p. 286. Another example is the root form **-cèng* (Guthrie CS 321), an old term for "cut," which forms the basis for words for "hoe" and "hatchet" in Vili, Kongo, Laali and Yaa, and Yaka, among others.

[41] Miracle, *Agriculture,* pp. 36–42. In the middle of the nineteenth century, for example, people in Tonga communities who had ivory and slaves to sell refused payment in cloth and beads, accepting only iron hoes. Their reasoning was based on their reliance at the time on the more difficult wooden hoes. Linda Heywood, "Production, Trade and Power—the Political Economy of Central Angola 1850–1930" (Ph.D. diss., Columbia University, 1984), pp. 126–27.

[42] See Reefe, "Societies of the Eastern Savanna," pp. 177, 190; and Vansina, "Peoples of the Forest," pp. 106, 109, and 115–16.

[43] Heywood cites several examples of slave owners who supplied their slaves with clothing and tools. Heywood, "Production, Trade and Power," pp. 96–97.

[44] The only cases I know of are the languages of Lengola (D12), Mituku (D13), and Bira (D32).

[45] For example, the blade on the right in Plate 452, page 794 of Johnston, *Grenfell*, Vol. II. Also, Kriger, museum notes, technical data for MRAC Acc. # 19100, 19115, 20876, 1796 2/1, 947, 948, 6454, 20882.

[46] Illustration 'Houe en fer du Lualaba' in "La Monnaie," *Le Congo Illustré* 1 (1892), p. 34. Also Kriger, museum notes, technical data from MRAC Acc. # 7270 2/1, 7271 2/1, 2/2.

[47] Johnston, *Grenfell*, Vol. II, p. 797.

[48] Reefe, *The Rainbow and the Kings*, pp. 95, 202. Reefe, who associates political control over iron deposits with formation of the Luba "empire," presents inconsistent evidence for tribute in iron. The limited data available indicate that tribute in iron consisted of refined bar iron and iron products, but not iron ore. The Munza iron district, located in the Luba heartland, was identified in oral tradition with the state's founder, Kalala Ilunga, but there have been no data gathered in Munza for estimating the extent and duration of mining and smelting there. Other important iron districts just outside the heartland, namely Kalulu and Kilulwe, are only briefly mentioned by Reefe. In neither of these cases does he cite any particular forms or amounts of iron tribute sent to the Luba court. His reference to tribute from another iron producing area, a place south of the heartland called Kapamay, contains no specific details. Reefe's only other example is from the village of Lubombo, whose blacksmiths claimed origins in Munza and sent tribute to the Luba kings, though again the form of this tribute is not specified. See Reefe, *The Rainbow and the Kings,* pp. 81–85, 89–90.

[49] Linguistic analysis may not help here, as there are few apparent sound shifts to trace. However, the languages belong to the K, L, M, and N Bantu language families, and the distribution is unbroken.

[50] P. Colle, *Les Baluba*, 2 Vols. (Brussels: de Wit, 1913), p. 789; and also the 1919 diary of Burton, in Moorhead, *Missionary Pioneering*, pp. 186, 195.

[51] Arrow blades were at one time a currency in the central basin, and served as an initial bridewealth payment. The Bobangi word for hoe, *lokóngo*, is shared among at least ten of the C group languages, sometimes in the form *loóngo* or *lóngo*.

[52] Kriger, museum notes, technical data from MRAC Acc. # 2158, collected before 1910.

[53] Joseph Clark, American Baptist Missionary Union missionary at Ikoko (Irebu) from 1894–1899, the hoes collected before 1903. FM Acc. # 33973 and 33974.

[54] Kriger, museum notes, technical data from Ntomba examples MRAC Acc. # 20844 and 20843, collected before 1917; and Uele examples MRAC Acc. # 2418 2/1, 2/2, collected before 1910.

[55] Kriger, museum notes, technical data from examples collected before 1890 by H. Chatelain, NMNH Acc. # 151310.

[56] See Vellut, "Notes sur le Lunda," pp. 72–73.

[57] Read, "Iron Smelting and Native Blacksmithing."

[58] Ibid., p. 48.

[59] United Church of Canada Archives, Victoria University, University of Toronto, Walter Currie Papers, CUR, File 1, letter Nov. 20, 1894.

[60] The regionwide lexical evidence indicates numerous such borrowings over time. This is so even for the root form *temo*, which shows several possible historical layers of borrowing. The one discussed here is one that appears to have been recent, with no sound shifts having occurred and the geographical distribution of the term being unbroken.

[61] Adriano Barbosa, *Dicionário Cokwe-Português* (Coimbra: Instituto de Antropologia, 1989).

[62] E. Fromont, "Souvenirs du Kwango et du Lunda St. Paul de Loanda" (1892–1893), MRAC, Section Historique, R. G. 1120, pp. 29–30. See also Barbosa, *Dicionário*; and Baumann, "Zur Morphologie," pp. 216–17.

[63] In the case of socketed spears, there was usually not enough iron in the haft of a blade to form a socket; but the socket of a socketed blade could be drawn down to make a haft.

[64] Jan Czekanowski, *Forschungen im Nil-Kongo-Zwischengebeit wissenschaftliche Ergebnisse der deutschen Zentral-Afrika Expedition, 1907–8*, 5 Vols. (Leipzig: Klinkhardt und Biermann, 1911–1927), Vol. II, pp. 31, 194–95; and Kriger, museum notes, technical data from MRAC Acc. # 36895, 29367, 2420, 2425, 6970, 6957, 2429, 11470, 11295.

[65] Lang Expedition (1909–1915), AMNH neg. # 224066.

[66] Accession information, MRAC Department of Ethnology.

[67] By cutting the hamstrings from behind to cripple and weaken the animal. This manner of hunting was usually done in small teams, and was most challenging and dangerous.

[68] C. van Overbergh and E. de Jonghe, *Les Mangbetu* (Brussels: de Wit, 1909), p. 477; Czekanowski, *Forschungen,* Vol. II, p. 133.

[69] Kriger, museum notes, technical data from MRAC Acc. # 19455, 19456, 2664 3/1 to 3/3, 11468 (6 specimens), and 20880.

[70] Kriger, museum notes, technical data from MRAC Acc. # 33700 2/1 and 2/2, 51.76.39 and .40.

[71] Rev. A.C. Good, American missionary on the Ogowé. Examples of the iron units are in the ethnology section of the NMNH, and his explanation filed under Acc. # 25654.

[72] Torday and Joyce, *Notes Ethnographiques* (1922), p. 86. See fn 9 for the objects I examined and weighed.

[73] Kriger, technical data from MRAC Acc. # 11470, and accession information.

[74] Kriger, museum notes, technical data from MRAC Acc. # 19455, 19456, 2664 3/1 to 3/3, 11468 (6 examples), and 20880.

[75] Kriger, museum notes, technical data from MRAC Acc. # 19499 15/1 to 15/15, 8889 2/1 and 2/2. Unfortunately, there is very little information available about these bar forms.

[76] Kriger, museum notes, technical data from MRAC Acc. # 14528 8/1 to 8/8 and other examples.

[77] See Thornton's brief analysis for the west coast of Africa in the seventeenth century, where he concludes that even with imported iron the local demand for iron could not be met. Thornton, *Africa and Africans*, pp. 45–48.

[78] For example in Kazembe in the 1860s, and Coquilhatville thirty and fifty years later. Joachim Monteiro, *Angola and the River Congo*, 2 Vols. (New York: Macmillan, 1876), Vol. II, p. 94; Charles Lemaire, "Une Forge à l'Equateur," *Le Congo Illustré* (1892); and A. Engels, "Les Wangata," *La Revue Congolaise* 1 (1910), p. 478.

[79] A blacksmith at San Salvador in 1880. W. Holman Bentley, *Pioneering on the Congo*, 2 Vols. (New York: Johnson Reprint Co., 1970. Reprint of 1900 edition.), Vol. I, p. 189.

[80] There have been few laboratory analyses done on iron objects from Africa that are in museum collections. For the Zaire basin, indications are that from at least the 1890s onward, overseas iron was used in manufacturing all sorts of products.

[81] See for example Bentley, *Pioneering*, Vol. II, p. 293; and Masui, *Guide,* p. 121.

5

The Work of the Finishing Forge

Kendá l'otúli lilóngó, kelá wene belemo bek'otúli by'olo.
Befriend a blacksmith and you can buy the better products.
—Mongo (Basankusu)[1]

Sendwe Banza wa kwibasambwila ne bityimukile ne bidi nsongo.
Sendwe Banza, the man able to draw out and repair even broken iron and make things sharp-pointed.
—praise name, Luba (Shaba)[2]

Economic expansion in nineteenth-century central Africa generated greater demand for iron, and as blacksmiths manufactured products to satisfy that demand, they engaged in important social business by making contacts with a range of other men. As we saw in the previous chapter, semifinished iron in the form of currency units often became the raw material for smiths who specialized in finished products, the circulation of iron thus linking blacksmiths with one another across space. This chapter focuses on a wider spectrum of blacksmiths' social relations, specifically how the design, production, and marketing of their products took them into different social worlds. I begin with a discussion of smithies and the blacksmith's tools in general, which leads into an analysis of hammers and anvils, masterworks that signified the founding of new workshops. Two sections are then devoted to examining the growth of luxury markets in the second half of the nineteenth century, as "big men" rose to prominence and accoutered themselves with fancy weapons and other prestige goods. In the final section, I propose that it was blacksmiths' strategies for keeping up with current market conditions, which involved inventing new products and retraining with masters to acquire new skills, that drove them to expand their horizons beyond their local communities.

Compared with smelters, the social relations of blacksmiths were, at least in an era of economic expansion, potentially far more diverse and wide-ranging. I argue that blacksmiths' identities were thus multifaceted and sometimes conflicting, that while their group identity was formed around gender and work, they sometimes rose so high in their occupation as to join elites, thereby transcending their residential and ethnic affiliations. I provide here selected historical examples that suggest such changes, based on my analyses of masterworks. What they reveal is the complex and multiethnic nature of blacksmiths' contacts and clientele, shattering any presumption that smiths worked mainly for local, subsistence production. In some cases, this social complexity was embodied in and revealed by a single product, one that emerged from a series of workshops in different locales. Moreover, by designing and marketing elaborate luxury products, master blacksmiths raised new standards of skill for their trade and at the same time entered the social networks and communities of ambitious "big men" of various sorts. I argue that iron was wealth and that ironworkers could and did enjoy upward social mobility—regionwide historical trends that become the focus for case studies in the next two chapters.

BLACKSMITHS' WORKSHOPS

Blacksmiths who specialized in finishing work usually set up their workshops at easily accessible locations in their communities, often at a well-traveled crossroads or near some other center of traffic. Smithies were in some rare instances enclosed and protected,[3] but for the most part they were only minimally sheltered from rain and sun, and were open to the view of passing onlookers. They were not simply workshops but were also meeting places, where people came to have their tools sharpened and repaired or to meet others, lingering to exchange news and information. The careful monitoring of outsiders, so evident in iron smelting, was not a priority here. Although there were some restrictions regarding nonspecialists' direct handling of the blacksmith's most important tools, hammers and anvils, these prohibitions were neither as stringent nor as elaborate as the systems of ritual that sometimes surrounded certain aspects of furnace construction and the smelting process.

Although blacksmiths sometimes claimed that they could unleash and direct supernatural forces, most of what went on in the smithy was skilled manual work and local gossip. Therefore, since there were so many more blacksmiths than smelters, studies that emphasize the importance of ironworking rituals convey an image of the craft that is dangerously skewed. For when we look at the occupation as a whole, including the blacksmith's workshop, it becomes apparent that ritual activities, important though they could be, did not account for the majority of ironworkers' time. Oral testimonies of smelters and reconstructions of smelts today may stress the roles of medicines and supernatural forces, but in contrast, the material and lexical evidence of blacksmiths' products stress the technical and management skills that were deployed at the forge. On balance, it is more accurate to characterize ironworking in general as productive and secular work.[4]

Smithing workshops were numerous and more evenly distributed geographically than were smelting furnaces, but they were neither ubiquitous nor uniform, and blacksmiths from place to place had different ways of organizing their product lines and volume of output. At one end of the spectrum, there were blacksmiths who only did repair work, and did not engage in the manufacture of products at all; at the other end were smiths who turned out various sorts of finished iron goods that were made specifically for sale in markets.[5] Part of a blacksmith's income also came from commission work, certain products made by special arrangement with individual customers; and there were some master smiths who worked for kings. Workshops that focused on producing in quantity only one single product, like the hoe-manufacturing center of Ndulo described in the previous chapter, were apparently not the norm. Instead, most blacksmiths seem to have avoided narrowly specializing their operations, doing repair and commission work as well as making a selection of product types that would find ready sale in whatever markets seemed promising to them at the time.

Management and training within the smithy thus posed ongoing challenges of different sorts. And just as the master smelter was the key to managing the processes of making iron, so, too, master blacksmiths managed and organized the manufacture of finished iron products. In certain respects, the factors of production and patterns of social relations that revolved around the forge were similar to the ones that revolved around the furnace. Like the master smelter, a master smith needed supplies of charcoal made from particular types of hardwood, and he might have to rely on other specialists for them. Like the smelter, he made his own tools and equipment. What labor he needed directly was mostly unskilled, supplied by his very young potential trainees who worked as bellows operators, and the other semiskilled apprentices in his shop.

A key factor that distinguished the social relations of the blacksmith from those of the smelter, at least in the nineteenth century, was the way he went about acquiring his supply of iron metal. He was usually dependent on outside sources for it, and had to contend with this problem on a continual basis. Moreover, there was always an imbalance among ironworkers in the demand for and supply of iron in their workshops, an imbalance that was created by seasonal constraints on the smelting and smithing production processes. Blacksmiths, unlike smelters, were not limited to working only during the dry season, but could work during the rainy season as well, when men's labor requirements were less burdensome. Thus the greater number of blacksmiths and their longer working season added significantly to demand for iron, and presented problems for them in keeping their workshops supplied.

In other words, iron changed hands in many directions, passing not just between producers and consumers but between ironworkers themselves. Some contacts were direct, between blacksmiths and nearby smelters, but indirect contacts were perhaps even more important. When individual customers or merchants supplied a blacksmith with scrap, bar, or semifinished iron, they were informing him, sometimes inadvertently, about the quality and type of iron products that smiths were making elsewhere. And similarly, when customers came in for repair work, or requested certain new products, or initiated a special commission, they turned smithies into centers of news about what was being made in other workshops. So in addition to organizing the labor and training within the workplace, master blacksmiths also had important social relations to manage. Certainly ironworkers developed commercial contacts for economic reasons, to supply themselves with iron and to sell their finished products. But markets were social worlds as well, offering blacksmiths opportunities for learning about new products, new skills, and potential avenues for advancement.

MAKING SMITHS AND SMITHIES: HAMMERS AND ANVILS

Smithies everywhere were built primarily around a set of portable tools, and the toolkit of a blacksmith was crucial to him in several ways. First, the extent and composition of his toolkit set certain limits on what kinds of iron goods he could produce. If he were limited to only certain types of large and heavy stone tools, he would be limited to refining iron, making bar iron units, and perhaps producing various kinds of small, simple blades and semifinished hoes or hatchets. A more extensive and varied toolkit obviously increased his options to engage in much more detailed and elaborate manufacturing work. Second, with a variety of tools and more of them a smith could bring in apprentices and divide labor in his workshop, thus increasing the range of his market share and the volume of his output. Third, by learning to make hammers and anvils, his most prized and useful tools, a smith demonstrated his mastery. The ability to perform this feat placed him in a position to regulate the craft by attracting apprentices and controlling the availability of essential capital equipment. In other words, blacksmiths' tools were used in the making of iron goods but above all they were used in the making of more blacksmiths and workshops.

Forges were open to the public and so were visited by some nineteenth- and early twentieth-century European and North American residents and travelers who described work in progress and the tools they saw being used. Often they were interested in particulars like energy supply and the degree to which work was divided or carried out cooperatively, as in this description of a smithy by a missionary as he traveled along the Aruwimi basin in 1899:

> At this place I saw one of the best native blacksmith shops I have met with. There were eight workers, but half of them were required for blowing the fire, by means of four small bellows arranged along one side of it. These bellows are of the immemorial pattern depicted in the ancient Egyptian sculpture pictures, and of the kind met with by travelers over the greater part of the continent. They are something similar to saucepans in shape, the wind being produced by rapidly working up and down baggy coverings of skin securely tied in the place of their covers, and finding its way to the fire through nozzles protruding from one side, just as saucepan handles do.
>
> The anvils are blocks of close-grained sandstone set on edge, and the hammers are pieces of iron in the shape of sharp pyramids, being about three times as high as they are wide on the face. The hammer handles are cleft sticks, through which the tapering hammer is inserted and made fast with split cane—a very insecure arrangement at first sight; but it appears to work well, the hammer being tightened up by the force of each successive blow.
>
> Each of the two smiths had two or three pieces of iron in the fire at the same time, and the one who was doing the heavier work—making an anklet—had two hammer men to help him; and as he had his own hammer as well, there were sometimes three of them going at the same time upon one piece of work; and wonderfully well they did it. Many of the pairs of anklets weigh seven pounds; those Lupu had on appeared to be considerably heavier than those I weighed.
>
> The blacksmiths had no tongs for taking hold of their work, but each piece of work had its own handle of green wood, by means of which it was held while being wrought. Very intricate patterns in fluting and chasing work were produced on knives and spears by means of a tool like a cold chisel set in a block of wood, over the point or edge of which the work was gradually moved, receiving a blow with a hammer on the reverse side at each movement, the pattern being provided on what is for the moment the underside of the work.
>
> As the people have no files, they very carefully hammer out their work to the exact size; the smoothing is then done with sand and water, and the polishing by carefully hammering with smooth-faced hammers. It did one good to see the industry of the people, for theirs was a day's work so much nearer the fair thing than is usual on the Congo, that it was really encouraging.[6]

Another visitor to this area seven years later considered the best forges to be the ones with a quadrangular-shaped iron anvil instead of or in addition to the more commonly used stone or cylindrical iron anvil. He neglected however, to explain fully the difference he understood this to make in the work. He may have recognized that some products, with very specific and detailed features, required tools with certain types of working edges and planes; thus, having a variety of hammers and anvils on hand was clearly advantageous. There were also other kinds of specialized tools that helped a smith to expand his markets. A very useful addition to smithies in the eastern Zaire basin was the burin, a tool that was key to the production of more elaborately embellished products. It was essential for engraving the linear surface patterns that distinguished many of the spear and knife blades sold along the Lualaba and Lomami Rivers.[7]

Toolkits and types of tools varied from smithy to smithy and from one general geographical area to another, though the evidence available is far from complete. Observers usually neglected to describe tools clearly and only rarely did they record their precise features in drawings or photographs. Some missionaries and ethnographers, realizing how important ironworking tools were, tried to collect them but usually failed because they were so highly prized. One missionary wrote from Bongandanga on the Lopori River in 1892 that he had bought a loom from a man he had seen weaving, but that when he met a blacksmith and tried to buy his tools the smith refused to sell them at any price.[8] Torday complained bitterly in 1908 that he had tried to buy an anvil from a blacksmith in the upper Lukenie River area but that all his offers were rejected. He also had wanted to buy a set of bellows, but had to travel for two days to find a smith who would sell one.[9] Nevertheless, some examples of blacksmithing tools do exist in museum collections, and when combined with other evidence they can show broad outlines of changes that took place in the work and workshops of blacksmiths over time, changes that reflect blacksmiths' efforts to expand their markets (Fig. 5-1).[10]

Hammers, for example, were of two very general types. In the north, northeast, and eastern portions of the Zaire basin, there were some blacksmiths who used hammers that were pyramidal-shaped iron blocks mounted in one way or another onto wooden handles.[11] Being more similar in design to the kind of hammer that European visitors were familiar with, these particular hammers were judged to be more advanced and were probably one reason why the forges where they were used were designated as most impressive. But to blacksmiths, these types of hammers were not necessarily the most useful ones, for although the handle offered the advantage of leverage it also imposed certain restrictions on the work that could be done with them. The various methods of mounting the iron head limited the working surface of the hammer to only one single flat plane. It is perhaps for that reason that these were not the hammers used by most blacksmiths in the region.

Most blacksmiths worked with hammers that were not mounted at all on handles but were grasped in the hand directly. Blacksmiths working in Maniema in 1874 had rope handles attached to their stone sledge hammers,

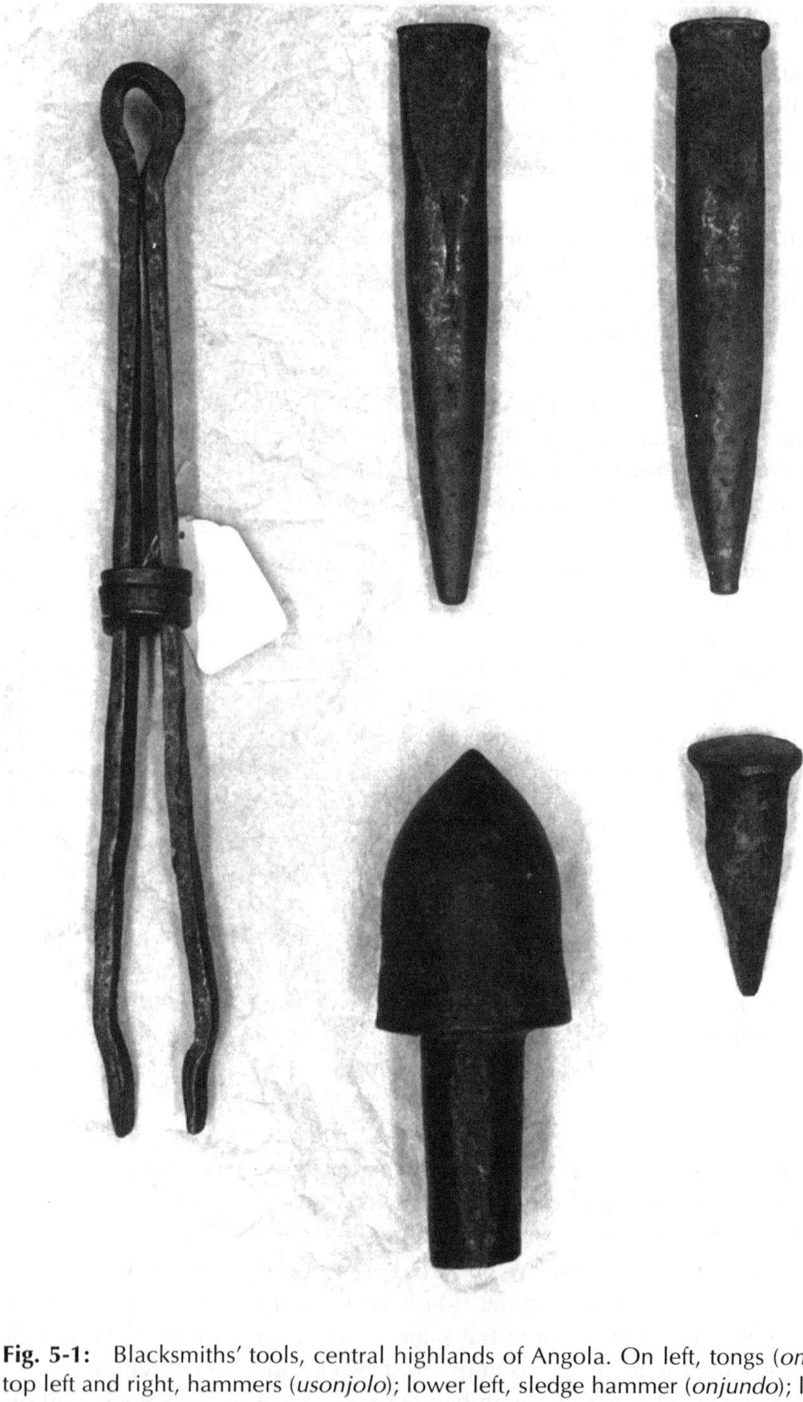

Fig. 5-1: Blacksmiths' tools, central highlands of Angola. On left, tongs (*omala*); top left and right, hammers (*usonjolo*); lower left, sledge hammer (*onjundo*); lower right, anvil (*olukata*). Not pictured is the stone anvil used with the sledge hammer. Courtesy of The Royal Ontario Museum, Toronto, Canada. HAC 84, HAC 86, HAC 88, HAC 87, HAC 85, respectively. Collected by Walter Currie, probably in the 1890s. [Photograph by the author]

while their smaller iron finishing hammers had no handles at all.[12] In the west, at Coquilhatville (now Mbandaka) along the middle Zaire, the toolkit used by a blacksmith there in the 1880s included an iron anvil and two different iron hammers, each without handles, as well as another set of equipment for working with brass currency units.[13] Just south near Lake Mayi Ndombe, a particularly well equipped and supplied smithy was inventoried by an ethnographer during the early rainy season of 1913. The master blacksmith in charge had managed to put together the following set of tools: a large stone serving as the "great anvil"; a brush for cleaning it; an iron anvil mounted on a wooden platform; three iron hammers (without handles) of different sizes and forms; a pair of double-drum bellows; a sharpening and polishing stone; a file; two forms of wooden tongs; several blooms of iron; and two containers of charcoal.[14] The various types of hammers that did not have wooden handles were more versatile than those that did, for they could be used in a wider range of smithing tasks. The blacksmith was free to use the hammer both laterally and vertically, pounding with its sides as well as one or both of its ends as he drew out and flattened the iron mass he worked on and shaped it into its final form.[15]

The most versatile hammer for finishing work was one that could be used laterally and vertically and also had two working ends. It was gripped in the center for pounding and held at one end or the other for lateral hammering. One such type was known as *bosákwá* in Mongo languages (see cover photograph), and many blacksmiths in the central basin included it in their set of tools. So did smiths in the Kuba kingdom, and it was probably through them that it was passed southward to blacksmiths working in Sala Mpasu communities.[16] Because this tool was used especially for articulating precise midrib forms and detailing complex blades, it is likely to have been eagerly adopted by smiths in the central basin, especially as markets for more elaborate products began to expand in the eighteenth century. It may have been Bobangi traders who were the agents in this process. About the inventor of it, however, and the subsequent master blacksmiths who perfected it, nothing is known—aside from their appreciation of flexible tools and working methods.

That this appreciation was widely shared among blacksmiths can be best illustrated by the so-called sledge hammer/anvil known as *nzunu* or *ndzùùnù* in Tio, for example, and as *njondo* in Mongo languages. Many different forms of this heavy blacksmith's tool existed throughout central and southern Africa, most of them known by lexical terms related to the same common Bantu root, indicating the antiquity of both the object and its name.[17] Used as a sledge hammer, it is usually associated with the making of hoe blades. The earliest form of the lexical term, along with some variant of the object itself, most likely was carried westward across south central Africa sometime during the first millennium C.E. as cereals were incorporated into western Bantu food production systems.[18] In west central Africa the tool was then adapted for use as an anvil for finish work, and in most languages of the rainforest areas the lexical term denotes either an anvil or a combined sledge hammer/anvil.

Although blacksmiths from time to time adjusted and redesigned the tool in slightly different ways, its versatility remained a constant. It could be a sledge at one time, and an anvil at another time during the production process. Twentieth-century blacksmiths continued to use it (see it being used as an anvil in cover photograph). An unusually attentive scholar observed a Pende blacksmith in the 1950s, noting especially the tools he used and the way he used them. The sledge/anvil played two important roles in his work. He began to make a lance blade by hammering out iron on a stone anvil, using the *nzundo* as a sledge. Then, for the finishing stages, he inserted the *nzundo* into a wooden base and used it as an anvil, along with a smaller hammer more appropriate for articulating the finer features.[19] It is not difficult to understand why so many smithies adopted this tool, for it offered convenience while widening the scope of a blacksmith's operations from heavy to detailed work.

While the flexibility and versatility of blacksmiths' tools demanded high degrees of skill in handling them, still other specialized skills were necessary to make them. One of the qualities that divided masters from ordinary blacksmiths and from mere apprentices was the ability to make a basic set of tools. Tools were loaned out during training, and it was the presentation of a hammer and sometimes an anvil that marked a man's graduation to the independent status of journeyman. But advancing to mastery involved further training in the expert work of making a set of tools. Highly skilled masters made bellows, either of wood or of clay, and guarded them vigorously. They did not manufacture stone anvils, but selected them, and since these tools were so essential for refining blooms and the rough work of consolidating masses of iron, well-chosen ones were highly treasured. Master smiths were the only ones who could make the large iron sledge hammer/anvils and the smaller anvils for finishing work. And the master smith made the various sorts of hammers that were necessary for producing particular types of products.

Training of smiths was rarely described in any detail, but the available accounts do demonstrate how the making of tools was integrated into the apprenticeship process:

> The making of the blacksmith's tools is a fetish, with which are connected many interesting rites. The sledge hammer takes a whole day to weld, and is done by experts; the receiver, who has served his apprenticeship, pays a large pig to the makers. Beer is brewed by the women of the village, a fowl is killed, and its blood sprinkled over the hammer after it is finished. It is then tied on the back of a young girl with a cloth, as a child is tied on

> by its mother, and then carried to the village of its owner, accompanied by the chief blacksmith, the others who worked with him, and their relations, and the whole village, singing special choruses for the occasion.[20]

Elsewhere in the region, east of the Lualaba River approaching Lake Edward, blacksmiths in the early twentieth century recounted how apprenticeship had usually progressed according to their experience. After serving as bellows operators during early childhood, an apprentice began more serious work by practicing how to heat a bar of iron until it was glowing red, and how to hammer it out. He began with small articles that were easy to make: small knives, and simple arrow and lance blades. He then moved up to objects considered more important, such as axes, hoes, and scythes or billhooks. Then he learned how to make a bellows, a crucible for melting copper alloy, and models for casting. Then at this time he would be designated a blacksmith. If he were the member of a chief's family, and if his father had been a master, he would also learn to make a hammer. And this was an operation embellished with ceremony. A goat and chickens were sacrificed and the ceremony, which lasted a day and a night, was done in secret with four or five masters working together, reputedly no one leaving the site for any reason. When the hammer was finally completed, it was heated to the highest temperature possible, then plunged into water to quench it. The master blacksmith who knew how to create a hammer passed this esoteric knowledge on to his chosen successor.[21]

These examples suggest a careful regulation of entry into and advancement in the occupation by controlling access to capital equipment and specific skills. Indeed, master blacksmiths were always in relatively short supply, and there are indications that at certain times and places they enjoyed monopolies over certain skills and products. Just as there were some master smelters who were itinerant at times or were invited to come and work in other communities or workshops, so, too, there were cases of master blacksmiths working on the road. In the Kasai district at the beginning of this century, master smiths were very much in demand in some locales and traveled regularly from workshop to workshop.[22] But how much and how consistently blacksmiths regulated access to tools and skills was apparently more variable than simple accounts of apprenticeship suggest.

Certainly the transfer of tools, and, by extension, the establishment of new smithies, was closely tied to the apprenticeship process. So, too, was the acquisition of skill. And general descriptions of apprenticeship stress quite consistently that blacksmiths trained their sons. Accounts also sometimes mention that the inheritance of a blacksmith's tools followed patrifilial patterns. What emerges is a seemingly clear image of control by smiths over entry into the occupation. Still, some questions linger about how much or how little ironworkers themselves were constrained by the norms of family and lineage relations that are embodied in these accounts. Monopoly control would imply that tools and skills would remain within the lineage, and, by extension, within the ethnic group. What appears to have happened, though, is something far more interesting, and it shows how blacksmiths' identities and relations did not completely conform to the established familial patterns of lineage ideology.

Accounts of apprenticeship that include any detail go on to state that non-kin and strangers could be accepted for training as well. In some cases, fathers were fathers in name only, training their slaves or the slaves of other "fathers."[23] Inheritance patterns, too, were variable and one finds that sorting them out into a simple set of alternatives is next to impossible. Sakata blacksmiths stated at the beginning of this century that a blacksmith's tools would go to the eldest son of his elder sister. But then they were able to offer so many secondary and tertiary options that no clear normative account could be given of exactly how a master's tools would be passed on after his death.[24] The recipient, however he was designated, was certainly fortunate, for it would be a boon to a young smith's income to inherit the tools that a master had accumulated over a lifetime. What is of greatest importance historically is who those young smiths might possibly be. Tools could be acquired directly from blacksmiths through training and also by inheritance, though the men who acquired them were not necessarily members of the family.

These normative accounts with their many exceptions and alternatives suggest that while master blacksmiths were able to regulate the numbers and locations of smithies, they were not at all rigid in doing so. They controlled the production of capital equipment, but they were not always completely restrictive. They appear to have been most careful within their own communities to maintain or modify their monopoly on the occupation and its skills in accordance with family and lineage ideology. Rituals, for example, called attention to the ways outsiders were restricted from training. At the local level, the making of hammers and anvils, and the ceremonial presentation of tools to the graduating apprentice, would have appropriately conveyed a sense of their importance to the continuation of the ironworking profession itself along family lines. Like the rituals of the iron smelt, rituals surrounding the manufacture and use of blacksmiths' tools amplified the master's control over the work, his powerful place in his own community, and his support of existing social and political hierarchies.

But in practice, as the histories of specific tools discussed above have indicated, tools and techniques were not rigidly controlled and could be transferred to ironworkers in other communities. Some masters, under certain conditions, even produced valuable smithing tools in surplus, to be traded beyond their own workshops. The most prominent example was a wave of production and trade of the sledge/anvil in the middle Zaire area, calculated by Dupré as having taken place during the seventeenth century.[25] It was during this time or before that iron smelting in the environs of the Téké plateaus was curtailed, leading to a more general transformation in iron production as master ironworkers dispersed, creating satellites of this former center. In prospecting for ore sources, they founded new smelting centers in locales inland and up tributaries on the right bank of the Zaire and along the lower Kasai, and set to work generating supplies and equipment for trade. Having these tools more widely available allowed some blacksmiths to retool and others to found new workshops.

In this same geographical area, a trade in blacksmithing tools continued into more recent times, though not on the same scale. During the late nineteenth and early twentieth centuries, sledge/anvils and finishing hammers were used for bridewealth and other sorts of payments and were probably also bought and sold as commodities. Meanwhile, master blacksmiths developed the sledge/anvil into a more clearly articulated form, one with a notched, mushroom-shaped head which gave it much greater stability when it was being used as an anvil. At least two variants of it circulated in the trade networks centered on Malebo Pool, along with the slender finishing hammer known in Tio as *mutyeno* or *-tyéné*.[26]

Given their importance in the refining of iron blooms, the making of bar iron and certain commodity currencies, as well as the decentralized organization of ironworking itself, the production and circulation of these tools is not so surprising. It is likely that during times of economic expansion, master blacksmiths who had reliable sources of iron would not view overproduction of capital equipment as a direct threat to their own livelihoods. To be sure, they would have known that, unless a man had the skills to work with them, the tools were valuable mainly as masses of metal, as stores of wealth in iron. But even the availability of those critical skills did not always follow the stated norms. Although they were very carefully guarded and managed, specialized skills were not completely confined within family relationships and bloodlines. Ironworkers were defined not by "lineage" or ethnicity or language but by gender—there were no women smelters or smiths, only men.

WORKING WITH HUNTERS AND WARRIORS

Blacksmiths had complex social relations with influential men, including contacts with other blacksmiths, with merchants, and especially with groups of important consumers. Among the latter were some of the men most involved in food production, hunters. All over the Zaire basin, men (and sometimes women) fished and trapped, bringing in crucial sources of protein for local consumption and for trade. Some techniques for hunting and fishing required iron weapons, and they came in all shapes and sizes, specially designed and adapted for killing certain kinds of animals in very specific ways. Designs for these weapons were not uniform throughout the region, but had been developed from place to place—in this way here or in that way there—as a result of ongoing collaboration over time between the blacksmith makers and the hunter or fisher users.

Also over time, these tools had become standardized in form and each recognized and prized for its particular efficacy. Blacksmiths made them and sold them not simply for the hunters and fishers in their own villages and towns, but for the ready markets further afield that were known to merchants. By the nineteenth century, certain examples of these tools were widely available, produced by blacksmiths of various language groups and traded over long distances. Several types of heavy harpoon used for killing large river mammals, one example with a detachable blade, were made in smithies all along the middle Zaire and its tributaries; blacksmiths further south, working in the upper Zambezi, upper Lualaba, and Luapula watersheds, made other kinds of highly specialized fishing spears.

Of course the biggest catch of all was the elephant, and well before the introduction of firearms various hunting methods and weapons had been developed for bringing one down. Two of the main methods requiring iron weapons were described in the 1880s. The first involved platforms which hunters constructed high in the trees along known elephant trails. Stationed in these platforms, hunters armed with heavy iron spears were positioned well to drive their weapons down between the shoulders of the beast as it passed below.[27] Blades for these spears had to be strong, sturdy, and well designed so as to pierce deeply enough into the elephant's flesh to seriously wound it. The most common one along the middle and upper Zaire and in the northern forests was a leaf-shaped blade with a broad and thick midrib down its center which added to it both strength and mass.[28] All over the Zaire basin region, wherever hunters used them, such elephant spears were among the largest, heaviest, and best-engineered blades that blacksmiths made.

The second method described for hunting elephant was much more daring and dangerous. Hunters armed only with a very broad, fanlike spear blade mounted on a long, sturdy handle, tracked the elephant on foot and sneaked up on it from behind. Once within striking distance, the hunter plunged the blade into its groin or cut one of its tendons, crippling it enough to be able to complete the kill.[29] This more solitary method was probably the oldest and certainly the most heroic of elephant hunting feats, and may originally have been developed by Twa hunters who then selectively taught the techniques to others.[30] It called for a broad-bladed weapon, in recent times an iron spear designed for cutting rather than for thrusting. At least two major standardized forms had been developed in the Zaire basin. One was the long, heavy hatchet-spear produced mainly by blacksmiths in the northern forests and mentioned in the previous chapter; the other one, made by blacksmiths in the environs of the middle Zaire and Malebo Pool, was fan-shaped and smaller in scale. The most common lexical term by which the latter blade was known shows a distribution suggesting that it was traded and used among western forest peoples over the past few centuries. Elephant hunters then had blacksmiths adapt it to be fired from guns in the late nineteenth century![31]

Along with the spear, arrows were essential weapons of the hunt. Iron arrow blades were common almost everywhere in the region, though noteworthy innovation in blade designs was not. In most cases, arrow blades were plain, leaf-shaped, and pointed, different types distinguished from one another by having one or two barbs or none at all. Much greater variety was possible, however, and there were at least two major centers in the region where hunters had worked closely with blacksmiths in developing novel and highly specialized iron arrow blades. In the lands stretching between the eastern borders of the Kuba kingdom and the middle Lualaba River, blacksmiths had perfected designs for at least fifteen different types of arrow blade. One had barbs, one had a protruding hook, one was jagged, other blades were quadrilateral or ovoid in form, or had several points, each type of blade also having its own individual name.[32] Although the specific rationales for each design may not be retrievable, it is clear that each one had been designed to perform in a prescribed way.

The other major area where arrow blades were highly specialized was around the headwaters of the Kwango and Kasai Rivers. There, hunters and blacksmiths from Cokwe, Ngangela, and Lwena communities (and probably others as well), developed at least eleven distinctively formed types of arrow blades for specific purposes in hunting small game. Hunters from this area were considered among the best in the Angolan interior, and certainly their hunting expertise owed a lot to the development of these weapons. Arrow blades were only one result of what must have been a long-standing collaboration between hunters and blacksmiths in this area. That collaboration continued during the second half of the nineteenth century as firearms were adopted increasingly for hunting elephant ivory. Cokwe blacksmiths achieved renown during that time for their skill in repairing and replacing parts of damaged firearms and mending split barrels.[33] How this expertise developed here and hardly anywhere else was rarely questioned, though such technical skills could not have arisen overnight. Surely blacksmiths' training in shaping such a variety of arrow blades with complex forms imparted to them a special aptitude for precision work. But it was only when they turned their attention to repairing firearms that this aptitude appeared most remarkable to outsiders.

In the oral traditions of Lunda, Luba, and other centralized polities, hunters were often credited with the founding of ruling dynasties, reflecting their powerful positions in central African societies and the respect they garnered from many quarters. One of the most well-known examples is Cibinda Ilunga, a hunter-hero of Luba origin and putative founder of the Lunda Mwata Yamvo dynasty. Carved figures representing him were made after the 1860s by pro-

Fig. 5-2: Hunter-hero Cibinda Ilunga. Figure carved in wood, with hair and fiber additions. 39 cm high. © Bildarchiv Preussischer Kulturbesitz, Berlin. Museum für Völkerkunde, Berlin, IIIC 1255. Collected by Otto Schütt, 1878-9 Schütt Expedition. [Photograph by R. Hietzge-Salisch]

fessional woodcarvers working in the service of Cokwe leaders, and he was always depicted with the paraphernalia considered necessary for a great hunter to carry (Fig. 5-2). And since it was elephant hunters equipped with firearms who were the most powerful and wealthy hunters at that time, Cibinda Ilunga figures usually carry guns, along with the small hatchets and knives they used for all sorts of incidental tasks.[34] Such heroic images, if they were ever carved before the introduction of firearms, would have portrayed a hunter carrying, instead of a gun, a bow and arrows, weapons that marked members of the society of professional hunters.[35] In either case, the presence of the smith is clearly implied in the imagery. By stressing the precise details of equipage, carvers drew attention to the dependency of these powerful and influential hunters on blacksmiths, the makers and repairers of their tools and weapons.

Hunters were not the only prominent and powerful male customers with whom blacksmiths collaborated. Weapons had been developed in specialized forms for war as well as for the hunt, and they were not all displaced by the increasing importation of firearms during the nineteenth century. Various models of European-made muskets and rifles did become coveted military weapons, and they were supplied to the Congo Free State's Force Publique after 1890, but they were not the only arms used in combat or for protection. Spears and arrows continued to be effective long-range weapons used by men in battles everywhere in the region, and they were especially lethal when combined with guerilla tactics. Near Stanley Falls, Force Publique troop columns learned this very quickly. There they suffered merciless attacks by spearmen who hid themselves in swamps, waiting to launch their weapons just when the soldiers were most vulnerable.[36] One officer in the Force Publique reported with glowing admiration on the weapons and tactics used by Twa ambushers. Extremely powerful poisons applied to their spear and arrow blades meant that even slight wounds could be deadly. And by timing their volleys carefully, archers were able to inflict serious losses on Force Publique expeditions, even though they lacked comparable firepower.[37] Some of the most dangerous arrows were made of wood only, with notches cut into the pointed end for holding poisons. Portions of these light, barbed missiles stayed in the wound, allowing these very lethal poisons to work more effectively, and as a result they caused far more casualties than guns did.[38]

The northern and eastern sections of the Zaire basin were places where blacksmiths had invented the greatest variety of spear blade designs. It is not an easy matter, though, to identify which ones were made specifically to be used in war. Although the spear or lance was known by a generic name common to languages throughout the Zaire basin region,[39] more specialized names had been given to specific types of spears. Most of these names referred to blades that had been designed for hunting certain game animals; some were made as currency blades; still others were used in war. But it is unlikely that there were rigidly held rules for which blades were used for which purposes, and some kinds of hunting blades could serve equally well for fighting. Those designated as war spears tended to have long, sharply pointed blades without barbs.[40] Generally speaking, since spears were conventionally used as long-range, airborne weapons, a blade that was sturdy though light in weight was most suitable as an offensive arm against human beings.

Spear blades and arrow blades changed hands often. They were sold and traded of course, and some were currencies in various places at times, but another way men acquired them was as trophies captured in war. More iron blades changed hands more quickly in the nineteenth century, not only because trade in iron products was increasing as economies expanded, but also because raiding and warfare increased. Surrender of arms, impoverishment and bankruptcy, theft or pillage by soldiers or colonial sentries, death of the owner by war or epidemic—these were some of the violent ways in which personal property was alienated and sold during the nineteenth century. Armies of slave raiders gave way to armies of colonial occupation, though there was little perceptible difference in the ways they operated. One factor that must have encouraged theft and pillage is that the lower ranks of the Force Publique, that is, almost all of the soldiers of African origin, were amongst the most poorly paid colonial forces in Africa.[41] One colonial officer described the consequences:

> When the patrols came in that evening we had quite a collection of native weapons to choose from—spears, shields, and knives. The shields—made of rattan and ingeniously wrought and sometimes quite ornamental, generally including a design in black, presumably the coat-of-arms of the warrior who bears it—we threw upon a fire and burned, while the spears and knives were retained by the soldiers who took them. Their ultimate destiny, no doubt, was to be sold to the white man passing up and down the river, who pay exorbitant prices for such articles. A total of sixty was the result of the day's hunt, and, as the proportion of spears was unusually large, all concerned were accordingly pleased. The capture of so many spears showed the presence of warriors in this region, and extra care was consequently taken.[42]

And as violence spread and intensified, so did peoples' need for another category of weapons, the kind necessary for personal, short-range defense. All along the northern and eastern Zaire basin, where slave raids became a commonplace after mid-century, blacksmiths invented various new defensive weapons out of spear blades. The economic and social conditions that brought more spear blades into circulation—as trade goods and as pillage—spawned rapid product innovations, taking already existing weapons and adapting them to serve new purposes. By remounting them onto shorter handles, blacksmiths quickly converted spear blades from offensive weapons into defensive knives or swords. Such innovations had taken place in the past, but they now became more numerous. Especially during the last quarter of the century, these particular types of personal weapons were being assembled in smithies along the Ubangi, Aruwimi, Lomami, and Lualaba Rivers—novel products composed out of the blades of older ones.

Some were made out of spear blades that were strong and relatively massive. One example that was made and sold in the north around the Uele basin as a sturdy, leaf-shaped spear blade for elephant hunting also circulated as a currency in the middle Zaire river trade.[43] In the Ubangi and Ngiri basins and southward, blacksmiths mounted it onto a short wooden handle and it could then be used as a knife, sometimes carried under the arm in a suspended sheath.[44] Their handles embellished with carved patterning, metal wire, and imported brass tacks, these knives are evidence also of the ever growing class of men who preferred their personal weapons to be both useful and visually impressive.

A better known example of a spear blade converted into a knife was carried by men traveling in slave and ivory caravans during the 1870s from the environs of Malebo Pool to ports on the northern Angolan coast.[45] Commonly referred to as a "war knife," it was also used inland, by river traders and other men all along the middle and upper Zaire as far as the mouth of the Lomami River.[46] Examples in museum collections vary, from those that have rather rudimentary, undecorated handles to others adorned with metal strip decoration, a range that suggests diverse origins for their former owners. They also show evidence of other additions. War knives were often taken to a specialist who would wrap protective charms and medicines into a package and attach it to the lower end of the handle.[47] That this was done regularly reveals quite graphically the degree of danger felt and experienced in the slave and ivory trades.

Further south, along the Lomami and Lualaba Rivers, several other types of these new personal weapons were made from very thin, light spear blade currencies combined with bar iron units. Since these spear blades were so light in weight, therefore lacking the mass desired in a good, short-range weapon, blacksmiths made up for this deficiency by adding stemmed iron currency bars to the ends of the handles as weights and as embellishment. One type was made with an elegantly formed and incised spear blade from the Aruwimi basin, mounted on a short handle and carried in an underarm sheath by traders and other well-heeled men along the lower Aruwimi and Lualaba Rivers.[48] Some of the pyramidal-shaped iron bar currencies hammered into the handles were enormous, perhaps in these cases serving more for ostentation than for purely practical purposes. Another type was made by blacksmiths further south, who assembled very light, thin spear blade currencies with stemmed cylinders of bar iron.[49] Still other types were assembled out of slightly sturdier lance blades, and the bar currency at the end of the handle varied from iron or copper batons, to used firearm cartridge casings, to iron or cast copper cylinders.[50]

All of these personal weapons of defense were made in response to increases in insecurity and violence over the nineteenth century, and they illustrate one production method exploited by many blacksmiths in the region, the method of assemblage. Resourceful smiths with good trade connections could quickly assemble these products out of partly or wholly preformed elements that were sold on the market. Like the grandiose hatchets made near Luluabourg and discussed in Chapter 4, they were made out of money, selected forms of iron commodity currencies that helped facilitate trade in the region. Certain types of spear blades could easily be converted into knife blades by mounting them onto a short handle made by a woodworker or by the blacksmith himself. If the blade was particularly light and thin in structure, adding a weight at the other end of the handle made it work more effectively as a knife. The final result was a product that, by the end of the nineteenth century, had become an established, standardized form of knife in spite of its seemingly *ad hoc* development. But this was not the only kind of personal weapon that was made for men to protect themselves, and assemblage was not the only way blacksmiths worked to manufacture weapons of defense.

Other weapons were made by methods more conventionally associated with smithing, that is, not by assembling parts but by hammering, modeling, and shaping the iron metal into complex structures. Blades made in this way were the result of much longer periods of design and development, that is, more concerted and sustained collaboration between smiths and their customers.

The most impressive, effective short-range weapons that blacksmiths had developed and produced in the Zaire basin region were the heavy war swords made in smithies near and around the upper Kasai and middle Lulua Rivers. And unlike the personal weapons assembled from spear blades, master blacksmiths invented these swords and perfected them over time specifically to be used by warriors in hand-to-hand combat. They were not developed from blades made originally for other uses. Each of the several different variants was massive, structurally sound, and well balanced, with double cutting edges. One type had a pointed leaf-shaped silhouette and so could be used as a cut and thrust blade, while the others were primarily designed as cutting weapons. Their blades have been described appropriately as "machines," for they were clearly built to get their deadly work done efficiently, and although their handles were sometimes adorned with carved and inlay decoration, the blades were always entirely, fiercely functional.[51]

It is not known precisely who the master blacksmiths were who first developed these swords, but it is possible to suggest where they worked based partly on evidence of where they were made and used during the nineteenth and early twentieth centuries. When examples of them were recorded or collected, they were described explicitly as war knives, though with different vernacular names attached to them: "muquale" for Luba-Kasai;[52] "tschiboschi" for South Kete;[53] and "goribi" for various ethnic subgroups within the Kuba kingdom.[54] When the recorded references to it are combined, a geographical area takes shape that includes portions of the Kuba kingdom and south of there toward the headwaters of the Lulua, lands bordered on either side by the Kasai and Lubilash rivers. The users of the sword were not confined to any single ethnic group, and the same can safely be said for the makers of it. Variations in certain formal details, and the different vernacular names of it, suggest that several distinct workshops developed the weapon over time, workshops operated by master blacksmiths most probably from Kete and Sala Mpasu, and perhaps also Ruund, Kanyok, and Luba-Kasai communities.

Controlling and wielding the weapons of war became the prerogative of the colonial state during the twentieth century, but instead of entirely disappearing, this sword continued to be used though in much more benign contexts. Referred to today as a ceremonial sword, it is now associated mainly with certain mask performances designed to restore collective harmony and order in Sala Mpasu communities.[55] But to call it simply a ceremonial sword, one would have to ignore completely its structure and design, and also its history, which together identify it as a superior form of defensive weapon.

That such an effective weapon was not made in innumerable smithies all over the savanna areas need not have been because it was invented only recently. On the contrary, the complexity of the blades' cross-sectional geometries and the coherence and integrity that guided their replication by different groups of smiths indicate that this is an object of some considerable age and pedigree. It may have been invented much earlier, but this weapon was likely to have been further developed and refined during the seventeenth and eighteenth centuries. These were times of upheaval and conflict, when the Kuba kingdom was engaged in wars of expansion and also when, beyond her southern frontier, political competition was intensifying between Ruund, Kanyok, Kete, and Sala Mpasu war leaders. Weapons production must have assumed greater importance than ever, along with master blacksmiths, the ones with special skills for manufacturing arms. Oral traditions confirm what one would expect, that during this period war leaders recruited specialized master smiths from both local and foreign workshops to provide high quality supplies of arrow, spear, and sword blades.[56] In this way, what surely would have been certain masters' monopolies on the manufacture of specific effective weapons, including this particular sword, were not kept solely within their own home communities or ethnic groups. At the same time, though, masters apparently were selective in how they transmitted certain specialized skills to blacksmiths in other locales.

It was access to and control over ironworkers' skills, rather than control over iron itself, that was so important politically. And much of the history of ironworkers is about how ironworkers managed to exploit this fact. Their ability to manufacture strategic weapons gave them leverage in local politics, as the following episodes from Sala Mpasu and southern Kete history illustrate.

Already mentioned was the importance of a class of warriors in the Kasai region, and how necessary to them were the weapons made by blacksmiths, especially the great war sword. If a warrior wished to secure reliable arms supplies, one strategy was to buy an apprenticeship for one of his male slaves. Master smiths agreeing to such an arrangement benefited from the gifts and payments in goods and foodstuffs made to them by the titular fathers of their apprentices. And the apprentice-slave was now in a more secure position himself, possessing such indispensable skills.[57] Sala Mpasu smiths who had been trained in this way as slaves tended to have lower social status than other smiths, their high degree of expertise trumped by the stigma of being unfree.

Whole communities could secure reliable arms supplies by encouraging blacksmiths to immigrate. This strategy was followed by the Ana Nzashi Kete, who sometime before the middle of the nineteenth century invited five Sala Mpasu blacksmiths and their families to come and settle in their territories, offering them land, security, and steady markets for their goods. These immigrant smiths established their homes just inside Kete lands, and as they grew wealthy they invested their profits in women and slaves. Their settlements became large, growing faster than other neighboring ones, and were known as cosmopolitan centers for the exchange of information and ideas. Ties with their home communities remained important all along, though, for it was there that they returned periodically to smelt or purchase their bloomery iron.

These same blacksmiths and their invaluable sword-making skills helped to drive events leading to the rise of a chief named Buanga who governed the Ana Nzashi Kete from the 1860s. His path to chiefship had been an unorthodox one, much of his influence coming not from the usual lineage or secret society connections but rather from his prowess as a warrior and his good relations with the blacksmiths' communities. These assets, and his successful investiture as a Kanyok chief, lent credibility to his offers of security and protection, allowing him to establish authority over Ana Nzashi communities and those of the smiths.[58]

In the process, however, the blacksmiths were forced to acknowledge and act on their own very divided loyalties. Members of their households and families were of mixed origins, and they themselves had strong ties to both the Sala Mpasu and Kete, some of these ties affirmed through marriage. Nevertheless, with the successful rise of Buanga, they chose to stay and thus

Fig. 5-3: Warrior with sword. Figure carved in wood, with some pigmentation added. 74 cm high. © Bildarchiv Preussischer Kulturbesitz, Berlin. Museum für Völkerkunde, Berlin, IIIC 3246. Collected by Hermann von Wissmann and Ludwig Wolf, 1884 von Wissmann Expedition. [Photograph by R. Hietzge-Salisch]

be incorporated into a new Kete polity. And so what originally had been a Kete request for access to blacksmiths' products ended in the transfer of the smiths and their skills into another society, another language group. The political dimensions of blacksmiths' control over their skills, especially the more strategic ones, thus led to social changes as well, and the transformation of ironworkers' own community affiliations.

Slowly, then, even the most strategic weapons came to be made in the smithies of different ethnic groups. Great Lulua warriors of the nineteenth century owned these massive, formidable war swords, for example, but it is uncertain how or from whom they acquired them. One superb specimen was held in the right hand of a powerful warrior depicted in carved wood, dating from before 1885 (Fig. 5-3).[59] He is an impressive figure, representing that class of men who rose to power and renown with the help of blacksmiths' skills. Like the hunter figure described earlier, he is shown properly accoutered with the essential tools of his profession. Warriors, like hunters, depended on blacksmiths for providing them with up-to-date weaponry and keeping it repaired—from arrow blades to swords to firearms. They also carried more humble iron implements, like the small hatchets and single-bladed knives suspended from their belts, which were useful in traveling, and in daily tasks. Like the successful hunter, triumphant warriors were venerated, portrayed in commemorative carvings carrying their secret medicines for healing and warding off bad fortune, and wearing the headgear and leopard skins associated with political leadership.[60] These are idealized images of exceptional and legendary men, composites of known historical figures, convincingly rendered with veristic detail. Great hunters and warriors, and those that aspired to be like them, were among the most powerful members of a blacksmith's clientele.

WORKING WITH LEADERS, LORDS, AND JUDGES

The social and political impacts of the growth in external trade during the nineteenth century—in slaves, ivory, and other products such as wax and rubber—were profound. More men were able to gain power, though their hold on it might be only momentary. Foreign merchants from afar offered wealth and opportunity to ambitious men of all sorts; certainly hunters and warriors quickly found ways to acquire riches and followings of dependents either on their own or through alliances with traders and political leaders. Some of those men who already had power expanded and consolidated it; some saw their power bases erode, crumble, and fall away, reforming anew around rising upstarts. Whole communities moved voluntarily or fled in resistance to raids or prestation; new social groups were created out of collectivities formed by slaves, refugees, and rebels. These kinds of upheavals continued during the Congo Free State regime, with colonial officers and mercenaries added as new agents of violence, coercion, and economic advantage. The purchasing power of all of these men grew, too, and certain blacksmiths profited.

Intense political competition expanded consumer markets for the kinds of luxury goods that had long been considered necessary for men in leadership positions. Although violence and insecurity increased and the pace of change accelerated, the nineteenth century was not a revolutionary one. True, leaders fell from power and new ones rose to power more calamitously and rapidly than ever before. But once established, the bases for sustained political authority and legitimation remained much the same. Claims to leadership, whether over defined territory or over groups of people, were maintained by effective administration and protection, that is, by using military strength, redistribution of goods, and institutions for resolving disputes and conflict. Some of the key implements for performing and legitimizing these social tasks were certain symbols of political authority and insignia of office, many of which were made of iron.

One of them is already relatively well known as a result of having been studied to some degree. Single and double clapperless iron bells were among the insignia of leadership in west, central, and southern African societies from as early as 1000 C.E. It was mainly their uses as tools of communication, instruments for conveying important messages—from war signals to public announcements—that gave them such high social and political value.[61] In most cases they were rather complex forms made of iron sheets pressure welded together at the sides, and so making them was usually the specialty of only certain blacksmiths who had acquired the proper and requisite training. Still, that did not prevent them from being made in quantity. Small versions of both single and double bells circulated as a form of currency around the Ubangi bend in the late nineteenth century, and so did the single iron sheets that were used to make them.[62] But smiths could not just simply acquire these flat sheets and easily assemble the bells out of them. To produce iron bells, blacksmiths had to learn particular techniques, forming the sheets presumably around a mold, and securing the welded seams firmly. This training was not readily available. Like the skills for making blacksmiths' tools, these skills were esoteric, and they were guarded carefully. Instruction was limited to very select local apprentices and to certain other more highly skilled ironworkers who lived much further away.[63]

Iron bells, and other iron objects that served as insignia, were generally set apart from more ordinary products of the smithy. In the hands of the elite men who used them they operated more like symbols than as signs of any single particular office, for they were put to a variety of practical and ceremonial uses. Their meanings were many and changeable rather than part of any fixed system. They also tended to be more elaborate, fancy, and technically difficult to make than most other metal products, special feats of blacksmithing skill that were worthy of the leaders who displayed them. Therefore, a smith would have to seek retraining from a master if he wanted to add one of these products to his repertoire. And during the nineteenth century, as more insignia were made to satisfy growing demand from the upwardly mobile, new, more exclusive examples had to be created as the preserve of the most well-connected and powerful few.

Blacksmiths supplied this growing luxury market by replicating certain insignia as prestige goods in greater quantities and in more workshops. Items especially in demand were weapons carried by men mainly for ceremonial and status purposes. Blacksmiths worked to make fancier, finer, more impressive examples for their most prestigious customers. But production for this market was far from uniform, and it is in the proliferation of fancy weapons over the century that one can see a striking trend in the cultural history of the region. Over most of the equatorial rainforests and the northern half of Angola, it was primarily special types of elaborate knives that were made; in areas stretching southeastward from the edge of the rainforests and southwestward into southern Angola, it was special types of elaborate hatchets. Along the Kasai and in central Angola these two cultural areas overlapped.

This cultural division illuminates the ways in which blacksmiths' products were deeply embedded in central African societies. For these two categories of luxury products grew from strong roots in the material world, in firmly held preferences for one or another form of all-purpose tool, the knife or the hatchet. In part of the region the knife was more important and resonant, and so that was the category of product within which fancy goods multiplied; in an almost entirely separate part of the region it was the hatchet that was more resonant, so that was the form in which fancy goods were made.[64] The full set of reasons for why the knife here and why the hatchet there may never be known, but that such a division existed speaks to another aspect of the design process, the specific cultural conditions and tastes that guided the innovations of blacksmiths who worked with their wealthy patrons in mind.

Probably the best known and widely seen image of a central African leader displaying a fancy knife is the engraved illustration of King Mbunza, a Mangbetu ruler from about 1868 to 1873.[65] The king holds in his right hand an elaborately embellished curved knife, a type called *nadjata* or *nagata*,[66] used generally by wealthy and important men to enhance their reputations at public meetings and ceremonies (Fig. 5-4). Some of the most beautifully made versions of these and similar knives were works of art made for chiefs and kings, masterpieces by blacksmiths serving the Mangbetu courts.[67] Although they were part of a very broad regionwide cultural phenomenon, that is, the creation of fancy show knives for wealthy men, their prototypes were local and very familiar. Analyzing the specific form of this type of knife reveals aspects of its history and uncovers the rationales followed by its blacksmith designers. In this case, as with others, it was not a completely new type of

Fig. 5-4: Luxury knife, public display in the *Mu* dance, part of a ritual for blessing the community. This knife is similar to the one held ceremonially at court by King Mbunza in 1870, and variants of it were displayed by elite men in Azande and Mangbetu communities during the late nineteenth century. Negatives/Transparencies #112182, Courtesy Department of Library Services, American Museum of Natural History. [Photograph taken in August of 1913 by Herbert Lang, American Museum of Natural History Congo Expedition, 1909-15]

knife and it was not an object chosen arbitrarily to refer to a man's wealth and prestige.

The Mangbetu ceremonial knife was an art form of the court that arose out of the countryside, a beautiful object based on an agricultural tool, the sickle, which also could double as a weapon. Agriculture was, of course, the foundation of the economy, and one of the most important staple crops was the plantain banana.[68] Blacksmiths in the Mangbetu kingdoms and environs made at least ten different types of knife, five of them being curved or sickle-shaped hand knives. These plain, often double-edged tools were used sometimes for clearing brush or harvesting grain, but above all they were designed for the trimming and cutting tasks of banana cultivation. It is reasonable to suppose, then, that many blacksmiths in this area learned to make the tool as part of their regular training. More skilled blacksmiths would have learned additional techniques for making the more complex versions of it, and select masters were the ones able to make the most elaborate specimens that were works of art. Forms of the sickle thus provide a visual gauge of social stratification, from the very plain and practical tools to finer ones, ending with the most spectacular examples of show knives that were associated with the Mangbetu rulers and elites.[69]

Even as an art object, though, it retained its basic tool form and features, taking on various types of embellishments that set it apart as special. It was the concave edge of the blade that was used most often in work and this edge was the one emphasized in ceremonial displays. Finer sickle-knives were carefully crafted with complex midrib and bevel structures, then blackened and polished for striking visual effect. Sometimes one or several round holes were punched into blades of higher quality knives. Other additions stressed its role as a weapon, such as the small protrusions along one or both sides of the knife that were for parrying but served also as visually beautiful details. The finest of the knives had cast copper or copper alloy filling the holes, or in some cases entire blades were made of copper alloy.[70] As a ceremonial object, the sickle-knife was truly impressive, and it made direct reference to an important source of wealth that everyone could recognize, the wealth of banana groves. Plantations of them were managed by the most powerful and richest men, men who could then afford to display with pride an implement far too sumptuous and valuable to be used.

These sickle-knives were particular to a geographical area that centered on the nineteenth-century Mangbetu kingdoms, but encompassed a wider territory including the middle and upper Uele and south to the Aruwimi and Ituri Rivers. This specific knife form was a cultural element that distinguished this geographical area from others in the Zaire basin, while linking it historically to the east. As an all-purpose agricultural tool, the sickle was not as commonly used in the Zaire basin proper, but was most important in east central Africa, especially in areas of more intensive banana cultivation around

Lakes Tanganyika, Victoria, and Malawi, and on eastward almost to the coast. Where sickles were used, however, they were not all alike. Designs of the tool varied, for example, from the hand-held Rwandan version, which was shaped rather like a hook, to other types with more gently curved blades, sometimes also with a socket for mounting the blade on a long pole. Aspects of design that characterized the Mangbetu sickle-knife were its specific shape, structure, and double edges, features that betray it as a distinct innovation of blacksmiths in that vicinity alone.[71] Just who these blacksmiths were, though, is open to question. It should not be assumed this was a Mangbetu invention, for as both tool and ceremonial object, it was used not just in Mangbetu communities but also in many others, by speakers of a number of different languages in this area.[72]

Another example from the knife culture is a particular type of sword known usually by the name *mukwale*. Some were practical, double-edged, defensive weapons, but most of them were more ornate with elaborate decoration on the face of the blade.[73] They became coveted symbols of prestige and wealth in areas serving the Angolan overseas trade, that is, the commercial networks linking Benguela with the Lunda kingdoms of central Africa. By the last quarter of the nineteenth century, local political leaders in Lunda, Cokwe, Lwena, Yaka, and Mbundu communities proudly displayed such swords on ceremonial occasions and as they conducted judicial proceedings.[74]

Although the *mukwale* swords shared general design features in common, slight variations among them suggest they were made not in one single place but by a number of blacksmiths who worked in different locales and who spoke different languages. They all shared the same basic sword design, one that was not as heavy and massive as the Kasai war sword discussed in the previous section, but whose structure resembled a sturdy lance blade. When manufacturing these particular swords, blacksmiths formed a central, sloping midrib, beveled edges on both sides, and drew the iron out into a gently curved leaf shape. What local variations there were consisted of differences in dimensions, especially in the width of the blade, and also in the type and degree of embellishment. Most frequently, blacksmiths hammered stamped dot and incised patterns onto the metal surfaces along the midrib, artfully arranging lines, lozenges, and scallops that imparted a degree of elegance and refinement to the visual effect of the blades.[75] Some blacksmiths in the Kasai area added contrasting inlay of copper wire to these surface patterns, a technique that was more commonly used on fancy hatchets.

Perhaps the most interesting local variant of this sword type was the one developed by blacksmiths serving the luxury markets of Yaka and Suku communities around the middle and upper Kwango River. It resembled the others in the way the blade was formed and decorated, in the type of sheath in which it was carried, and in the handle decorations, which were made partly from the metal cases of discharged firearm cartridges.[76] How it differed, though, was in the final dimensions and proportions of the blade—in this case a silhouette much narrower and longer than others. The difference can be explained not as a deliberate effort to change the design but as the result of the different iron sources used by the blacksmiths who manufactured it. Although there were sources of locally produced iron available (such as the *kimburi* currency bars), smiths chose instead to recycle iron from overseas, that is, to make swords out of guns. Old firearms that were no longer serviceable were stripped down so that the iron barrels could be heated, hammered, and shaped into elegant swords.[77] In doing this, blacksmiths adapted their working methods to these unusually long, slim bars of iron, subtly changing the form and character of the blade in the process.

Along the southern margins of the Zaire basin region, the luxury market for male prestige goods included elaborately made and embellished hatchets. The relatively high social value placed on these products was similar to that for fancy knives and swords, and was based on how and by whom they were used. In tracking individual examples of them, one finds that their potential uses as weapons were augmented by important social uses: these were objects wielded both as tools and as symbols of status in political and judicial processes, legitimizing one's position and authority. Titleholders, diviners, and other notables proudly used and displayed them to inspire admiration and respect among their clients and followers.

A center for the manufacture and commercial marketing of ceremonial hatchets was near the colonial post of Luluabourg after 1885 (mentioned in Chapter 4). The manufacturers were the famous "Zappo Zap" [sic] blacksmiths, followers of the Ben Ekie (Songye) chief Nsapo Nsapo, who had fled to this new location about 200 km west of their homelands to escape the suzerainties and prestations of the Zanzibari merchant-sultan Tippu Tip and his allies. Praised by the European officials and mercenaries who lived near their workshops, these blacksmiths turned out impressive quantities of iron goods, including hundreds of fancy hatchets in at least five standard forms.[78] Some observers considered the Nsapo masters the best blacksmiths in all of central Africa.

What they excelled at was volume production and marketing. Their sources of iron were probably variable, from bar iron units made from local bloomery iron to the hoop and bar iron imported from overseas, and they sold their products to a varied clientele. Merchants bought the renowned elaborate hatchets, for example, which then found their way along trade routes to places as far away as Coquilhatville (Mbandaka), about 1300 km by boat from Luluabourg.[79] Many foreigners bought them, including missionaries and ethnographers, and as a result, many of these specimens of central African industry ended up in museums in Europe and North America. However, it would be a mistake to assume that they were made as curios for the tourist market. They were showy symbols of wealth, parade hatchets called *kilonda* in the

Songye language, and were eagerly acquired by powerful warriors and title-holders to be displayed on public occasions. And like the men who owned them, these objects of prestige could end up in faraway places. Two hatchets that may have been made in the Nsapo workshops were found at the site of Khami, though not in a stratigraphic context, over 1600 km south-southeast of Luluabourg.[80]

Production of these luxury goods was partly a response by certain groups of smiths to the increase in the number of men who were able to buy them. But their designs were variations based on two older, more prestigious types of hatchet that were not so rapidly produced or easily acquired. The Nsapo blacksmiths were not the most skilled at their trade, but they were certainly the most well known, especially to outsiders. Much more careful, detailed, highly skilled work went into the making of the hatchets that had served as their design models.

One of these older hatchet types featured beautifully modeled and chiseled miniature male heads of iron, placed in open areas of the hatchet blade like a face in a window. Very few examples of this type exist in museum collections, but they were distinctly different from the hatchets made in Nsapo workshops.[81] They were made as works of art. The handling of the iron is far more impressive, the sculpting of the heads entirely volumetric and more detailed than the rather cursory suggestions made in relief on the market-oriented products. The figural conventions of these iron artworks followed those exhibited in wood sculpture, the personal and collective divination statuary made by Songye carvers.[82] Another parallel was the use of copper inlay. One hatchet had a line of copper added to the head, dividing the forehead in half vertically, and continuing down along the bridge of the nose, a visual convention also made in sheet copper on Songye carved figures. Torday, the collector of one such unusual hatchet, called it an "old Basongye axe, not to be confused with the state axes made by the Zappo Zaps."[83] Little else is known about it, but such works were clearly not produced in quantity, and probably had been reserved for select men who held political or religious authority.

The other older hatchet type was not nearly as rare. It had been noticed by European explorers in the Kasai region in the 1870s and 80s, described variously by them as a "weapon" of the Songye, Kusu, and eastern Luba (Fig. 5-5).[84] It, too, was made with care, though examples of it differed from one another in the degree of elaboration in the structure and embellishment of the blade and decorations on the handle. Indeed, what is most striking about these hatchets is the way they seem to have been built up over time, by the accretion of more elements. Blacksmiths apparently had made additions and modifications to the object's design in order to increase its visual splendor. A center for making them was the Songye chiefdom of Kalebwe, though they were not produced in anything like the quantities of hatchets that came out of Nsapo workshops.[85]

Fig. 5-5: Luxury hatchet, owned and displayed by elite Songye, Luba, and Lunda men in Kasai and Shaba communities. Iron blade with openwork and copper inlay. Courtesy of The Royal Ontario Museum, Toronto, Canada. HAC 572. Collected by Walter Currie, ca. 1903. [Photograph by the author]

Examples ranged from the simple to the very fancy and there were many intermediate stages of elaboration in between. At one end of the scale was the simple iron hatchet blade inserted into a plain wooden handle, an implement that would have been serviceable as both a tool and a weapon. Embellishments added to this basic blade form included patterns of small punched holes or cut rectangular openings. Later, to make the hatchet much more impressive, it could be taken to another smith who specialized in the techniques of inlaying copper on iron. The blacksmiths who had these skills employed them in all sorts of ways, from creating copper medallions inside the openings in the blades to arranging linear patterns in copper such as loops and interlaces on the iron surfaces. Fancy additions to the handle were coverings of monitor lizard skin or layers of copper sheet, sometimes studded with the precious beads used as currency in foreign trade. Examples of hatchets with some or all these elements, though in various combinations, were made by nineteenth century blacksmiths, and it is likely that in at least some of these examples, different workshops contributed at different times to what became the final product.[86]

All of these ceremonial hatchets were held in high regard, some much more than others. But even the ones that were available for purchase in markets were socially valuable, some well beyond their economic worth. When the chief Nsapo Nsapo died in 1894, he was glorified at his burial with wrappings of cloth currency, leopard skins, his personal firearm, lances, and several special hatchets.[87] By the middle of the twentieth century, new status symbols were in vogue and new symbols of political authority were in place. Hatchets such as these were no longer made. Rare examples were kept, however, as treasures, time having turned them into potent symbols of dignity, objects of precolonial collective identity and pride. Some artfully made ceremonial hatchets and adzes remaining in the possession of chiefs and elders could not be bought at any price. In the 1950s, a Songye chief refused the offer of a car in exchange for his superb adze, saying, "If I sell it, I sell my whole village, and then I die."[88]

BLACKSMITHS' STRATEGIES

Each one of the iron masterworks analyzed and discussed above contributes to a new body of evidence for blacksmithing as it was performed throughout the Zaire basin region. Taken as a whole, this evidence then places in proper perspective the role of rituals in the craft tradition, showing them to have been sporadic and rather minor episodes compared to the ongoing work of making iron, currencies, tools, and a host of other products, many of them technically challenging. Rituals that consecrated ironworkers' tools and workplaces and the prohibitions that they enforced to insure success, explain failure, and monitor access to skills were important, to be sure. All of these activities do indeed call attention to the efforts by smiths to control and conserve their craft tradition. Nevertheless, special rituals should not serve as distractions from the ongoing, daily activities of the craft, namely the multifarious tasks that were involved in managing and operating a workshop over a lifetime.

Focusing on this new evidence from masterworks takes us deep inside a variety of working processes and offers up valuable inferences about the workers and how they organized themselves. Masterworks made during the nineteenth century reflect the economic expansion going on at the time, their many complex and standardized forms the material realization of the highly specialized skills of smiths and their drive for innovation. As central African economies grew, more iron products of all kinds were made and purchased, some for economically productive work, others for protection and warfare, still others for prestige purposes. The range of standardized iron products grew wider. And the histories of these products demonstrate clearly that blacksmiths were innovators, replacing iron imports from other parts of the region by learning how to make them and in some cases inventing entirely new forms of iron goods. Such innovation was encouraged by the patronage of wealthy patriarchs and merchants, men who sought luxury products made of metal that could verify their important achievements and mark how high they had risen in society. By the end of the century, the purchase of iron commodities and the conspicuous consumption of fancy iron goods probably reached its zenith. And throughout this time, demand for iron continued to outstrip supply, notwithstanding the absorption of imported bar iron from overseas into the production, currency, and trading systems. It is not unreasonable to suggest, therefore, that blacksmiths deliberately chose to diversify their output rather than focus on manufacturing a single product. Diversification would have been a pragmatic strategy to follow because the markets for ironwares were multilayered, multifaceted, variable in scope, and changing, sometimes rapidly.

Diversification made sense also because it was rare for these finishing workshops to have only one active blacksmith. Usually there were several smiths and apprentices working together under the same roof, making products appropriate to their different levels of skill and expertise. It was usual also for a young man's initial apprenticeship to extend over a period of at least several years, with instruction carried out on the job, by observation of other smiths and by working with and handling different tools. Until the time when he was designated an independent craftsman, the product of his labor belonged to the master smith.[89] Then, upon completion of his general training, a smith might have had reason to continue to work alongside his master, for example, or he might have chosen to leave and establish a new settlement around his own separate workshop. Such was the case around Lake Mayi Ndombe,[90] where many towns had been founded by a blacksmith who had set out on his own with his family and other followers. In so doing, they became the leaders and notables of new settlements and thus over time the number and locations of smithies increased.

Some blacksmiths, through their own persistence and canny enterprise, attained full mastery of their craft. And as masters, their special domain was manufacture of iron products that assured the continuation of their own occupation—smithing tools—and that augmented the authority of political leaders in society—weapons and insignia. Talent and skill were essential features necessary to attain the highest levels of mastery, but equally crucial was the ongoing work of networking and retraining, that is, the business of establishing contacts of all sorts, keeping informed about new products, and learning new techniques.

It was above all the production of masterworks that could establish the reputation of a master blacksmith, since masterworks—the most important smithing tools and leadership insignia—were the most challenging and difficult objects to make. Their invention involved collaboration between the blacksmith and some of the most important and powerful men around him. And once invented, masterworks could not be replicated easily from a prototype acquired through trade; instead, they required additional technical observation, instruction, and practice. Masters were not made during the initial apprenticeship period of their youth, but over years of work and keeping in tune with changes in what were deemed at the moment to be the most profitable product lines. Only after stints of retraining with master blacksmiths who were willing to pass on their most valuable skills, the ones necessary for making current masterworks, did blacksmiths advance. And as the masters' tools and society's fashions changed, so did the measure of masters.

NOTES

[1] Korse, "La Forge," pp. 32–33.

[2] Harold Womersley, *Legends and History of the Luba* (Los Angeles: Crossroads Press, 1984), p. 5.

[3] An example would be the smiths of Sala Mpasu communities.

[4] As opposed to those who would exoticize ironworkers. See Eliade, *The Forge and the Crucible*.

[5] In 1905, Frobenius noted Pende smiths near the mouth of the Tshikapa who only repaired iron goods; their counterparts upriver were manufacturers. Klein, *Frobenius,* Vol. II, p. 61.

[6] Letter written by George Grenfell, November 1899, published in Hawker, *Life of Grenfell*, pp. 441–42.

[7] A. de Calonne-Beaufaict, *Les Ababua* (Brussels: Polleunis et Ceuterick, 1909), pp. 47–49; Védy, "Les Ababuas," pp. 272–73.

[8] Letter from Mr. Whytock, *Regions Beyond* 13 (1892), pp. 77–78.

[9] Emil Torday, writing from Lodja, Aug. 9, 1908. Field notes, Torday Congo Expedition, MS "Notes and Drawings of the Congo Expedition 1905–8," large format volume, p. 85.

[10] Circumstances of how tools made their way into collections varied. Some may have been available for purchase from merchants, though many tools in collections show traces of use. Others may have been available for purchase after having been alienated from their owners during violent episodes of slave raiding or colonial forced labor/terrorism. For example, during the period of forced rubber gathering in the ABIR Company territory, a sentry apparently killed a blacksmith who had been fleeing to the forest, and confiscated his property. The missionary who reported this incident had seen the blacksmith's goods stored at the home of the sentry. Rev. John Harris, *The Story of the Bongwonga Rubber Collectors* (London: Regions Beyond Missionary Union, 1905), p. 18. The most extensive collection of tools is in the MRAC, and has been the subject of a study published by Maquet, *Outils de Forge*.

[11] For a general idea of where these types of hammers were used, see Map #1 in Maquet, *Outils de Forge*.

[12] Cameron, *Across Africa*, Vol. I, p. 372.

[13] Charles Lemaire, "Une Forge à l'Equateur," *Le Congo Illustré* (1892); also reprinted in idem, *Congo et Belgique* (Brussels: Bulens, 1894), p. 134.

[14] Joseph Maes, 9 Sept. 1913, Journal de Route, Archives Africaines, D61 (3853) No. 9, Vol. II, pp. 14–15.

[15] Personal observation, Lopanzo, 1989.

[16] The term comes from the verb stem *-sák-*, to hit, which is not a common Bantu form and was probably an innovation within the Lomongo group (C61). USalampasu (K21) shows the same term for a hammer very similar to the Kuba one. See Maquet, *Outils de Forge,* p. 40.

[17] The root **-jùndò* is variously defined as hammer, anvil, or hammer/anvil.

[18] This postulation is based on lexical data from close to one hundred Bantu languages, preliminary analyses of sound shifts in each term noted, and discussions with Prof. Jan Vansina, 1992–3. On the introduction of cereals into western Bantu agricultural systems, see Vansina, *Paths*, p. 66.

[19] See Maquet, *Outils de Forge*, p. 42. In my observations in Equateur, I never saw the tool used as a sledge, but that may have been because none of the blacksmiths I watched started their work with large masses of iron.

[20] Read, "Iron Smelting and Native Blacksmithing," p. 48.

[21] Borgerhoff, "Les Industries des WaNande," *La Revue Congolaise* 3 (1912–3), p. 281.

[22] This phenomenon was reported by Leo Frobenius during his travels in 1905–6 in Kasai. Klein, *Frobenius*, Vol. IV, p. 98. Verner also reported a smith having been asked by a political leader just outside the borders of the Kuba kingdom to come and work in his town. S. Verner, *Pioneering in Central Africa* (Richmond: Presbyterian Committee of Publication, 1903), pp. 201–203.

[23] See, for example, William Pruitt, Jr., "An Independent People: A History of the Sala Mpasu of Zaire and their Neighbors" (Ph.D. diss., Northwestern University, 1973), pp. 149–50.

[24] See Maes, "La Métallurgie."

[25] See M.-C. Dupré, "Pour une histoire des productions."

[26] G. Dupré, *Un Ordre et sa Destruction*, pp. 136, 393 fn 52; P. Timmermans, "Voyage à travers le Gabon et le sud-ouest du Congo-Brazzaville," *Africa-Tervuren* 10, 3 (1964), pp. 72–73; and Collomb, "Quelques aspects techniques de la forge," p. 54. The lexical term for this hammer was borrowed into at least twelve contiguous languages from three different Bantu language groups—a phenomenon that signals recent trade of

the object, though linguistic analysis of sound shifts will probably reveal several historical dimensions to this process.

[27] Herbert Ward, *Among Congo Cannibals* (New York: Negro Universities Press, 1969. Reprint of 1891 edition), pp. 122–23.

[28] See examples illustrated in Junker, *Travels in Africa*, Vol. I, p. 331, # 2; and Leo Frobenius, "Afrikanische Messer," *Prometheus* 620 (1901), p. 754, fig. 608.

[29] Ward, *Among Congo Cannibals,* p. 123.

[30] It was practiced in many parts of central and southern Africa. See, for example, H. C. Woodhouse, "Elephant Hunting by Hamstringing Depicted in the Rock Paintings of Southern Africa," *South African Journal of Science* 72, 6 (June 1976).

[31] See Joseph Cornet, "La Société des Chasseurs d'Elephants chez les Ipanga," *Annales Aequatoria* 1 (1980), p. 243.

[32] J. Hagendorens, *Dictionnaire Otetela-Français* (Bandundu: CEEBA, 1975).

[33] Klein, *Frobenius*, Vol. II, pp. 42, 72. See also Miller, "Cokwe Trade and Conquest."

[34] See, for example, Marie-Louise Bastin, "Tshibinda Ilunga: A Propos d'une statuette de chasseur ramenée par Otto H. Schütt en 1880," *Baessler-Archiv*, Neue Folge, 13 (1965); and idem, *Statuettes Tshokwe du Héros Civilisateur 'Tshibinda Ilunga'* (Arnouville: Arts d'Afrique Noire, 1978), pp. 104–107.

[35] See Miller, *Kings and Kinsmen*, pp. 127–28. Miller refers to the dagger, bow, and hatchet as their weapons, based on recent Cokwe accounts. The southern Lunda associate the hunters' society with bow hunting.

[36] Soldiers raised their firearms above their heads to keep them dry, thus becoming defenseless targets. L. H. Gann and Peter Duignan, *The Rulers of Belgian Africa 1884–1914* (Princeton: Princeton University Press, 1979), pp. 53–54.

[37] Sidney L. Hinde, *The Fall of the Congo Arabs* (New York: Negro Universities Press, 1969. Reprint of 1897 edition.), pp. 83–85.

[38] See the entries written about the "Bena Luidi" (Kasai) in the *Carnet de Voyage, 1894* by E. Fromont, MRAC, Section Historique, RG 1120, no pagination.

[39] See the discussion of the Bantu root **-gongá* in Vansina, *Paths*, p. 283.

[40] See, for example, the lances pictured and described in Franz Thonner, *Dans La Grande Forêt de l'Afrique Centrale* (Brussels: Société Belge de Librairie, 1899), Plate 84 and p. 101; and Johnston, *Grenfell*, Vol. II, p. 773.

[41] Gann and Duignan, *The Rulers of Belgian Africa,* p. 82.

[42] E. Canisius in Guy Burrows, *The Curse of Central Africa* (London: Everette, 1903), p. 121.

[43] Goad (Hope Morgan) Collection, ROM Acc. # HA 2484 and accession information from the collector, which is most reliable in this case as the object was probably acquired from Joseph Clark, a missionary at Ikoko who was attentive about matters of local crafts. Some examples of the same type of blade in the MRAC show no midrib, and had holes placed vertically in the center of the blade, suggesting they operated as a currency as Hope Morgan stated.

[44] See Coquilhat, *Sur le Haut-Congo*, p. 241; E. J. Glave, *Six Years of Adventure in Congo-land* (London: Sampson Low, Marston, and Co., 1893), p. 213; and a specimen collected by the Duke of Mecklenberg Expedition (1910–11), now in the Frankfurt Museum, Acc. # N.S. 17. 460 a,b, published as catalogue no. 85 in Johanna Agthe and Karin Strauss, *Waffen aus Zentral-Afrika* (Frankfurt: Museum für Völkerkunde, 1985).

[45] Monteiro, *Angola and the River Congo*, Vol. I, drawing facing page 145.

[46] See Coquilhat, *Sur le Haut-Congo,* p. 241; Thonner, *Dans la Grande Forêt*, Plate 85; Johnston, *Grenfell*, Vol. II, pp. 694, 740; idem, *The River Congo from its Mouth to*

Bolobo (Detroit: Negro University Press, 1970. Reprint of 1884 edition.), p. 434; and James Jameson, *The Story of the Rear Column of the Emin Pasha Relief Expedition* (Toronto: Rose, 1891), p. 201.

[47] Based mainly on my survey of over 75 specimens in the MRAC, Tervuren, Belgium, in 1989.

[48] Kriger, museum notes, technical data from the W. H. Leslie Collection, FM Acc. # 29960, 29962, 29963 and accession files; and from the Joseph Clark Collection, FM Acc. # 33926, 33927, 33928 and accession files. See also Johnston, *Grenfell*, Vol. II, pp. 577, 778; and E. Torday and T. A. Joyce, *Notes Ethnographiques sur les Peuples communément appelés Bakuba, ainsi que sur les Peuplades Apparantées, les Bushongo* (Tervuren: Musée du Congo Belge, 1910), p. 215.

[49] Kriger, museum notes, technical data from MOM Acc. # 1909.Ty.555, .556, and .557. I also examined about eighty specimens in the ethnology collections of the MRAC in Tervuren.

[50] Kriger, museum notes, technical data from the Goad (Hope Morgan) Collection in the ROM, Toronto, Accession # HA 2463, HA 2464, and HA 2465. I surveyed additional examples of these three knife types in the MRAC: about fifty, twenty-five, and a hundred twenty, respectively.

[51] Interview including J. E. Rehder, David Norrie, and Daniel Kerem, ROM, 10 April 1989. Description of the blades is derived from the approximately eighty examples I have examined in museum collections, as well as published and unpublished drawings and photographs. The knives have been grouped together despite their different silhouettes because they exhibit distinctive technical similarities, which link them together. This morphological connection strongly suggests historical ties among blacksmiths' workshops of the different language groups.

[52] Hermann von Wissmann, Ludwig Wolf, Kurt von François, and Hans Mueller, *Im innern Afrikas* (Nendeln: Kraus Reprint, 1974. Reprint of 1888 edition.), p. 217.

[53] Klein, *Frobenius*, Vol. III, pp. 39, 160.

[54] Kriger, museum notes, technical data from the Torday Collection, MOM Acc. # 1907.5-28.359, 1908.Ty.239, 1909.5-13.186, .187, .188, 1909.Ty.666, .668. The correct Bushoong term is *ngodipy*.

[55] See the series of photographs taken by C. Lamote for Inforcongo in the 1950s, on file in the Ethnology Section, MRAC. The sword was referred to as *muela wa nvita* or *mpuku* in Sala Mpasu communities. Pruitt, *An Independent People,* p. 149; MRAC, Ethnology Section, photograph taken in 1987 by R. Ceyssens, EPH 14408 and documentation.

[56] See John C. Yoder, *The Kanyok of Zaire* (Cambridge: Cambridge University Press, 1992), pp. 29, 32–37, 94–95; Vansina, *Children of Woot*, pp. 160–65.

[57] Pruitt, *An Independent People,* pp. 149–50.

[58] Ibid., pp. 283–89.

[59] Kurt Krieger, *Westafrikanische Plastik*, 3 Vols. (Berlin: Museum für Völkerkunde, 1965), Vol. 1, Plate 254 and pp. 119–20; Hans-Joachim Koloss, *Art of Central Africa* (New York: Metropolitan Museum of Art, 1990), pp. 58–59. In the latter text, the name of the figure is erroneously quoted—it should be *makabu buanga*, not as printed.

[60] M.-L. Bastin, *Statuettes Tshokwe*, p. 106; Koloss, *Art of Central Africa,* pp. 50, 58.

[61] Jan Vansina, "The Bells of Kings," *JAH* 10, 2 (1969).

[62] Kriger, museum notes, technical data from MRAC, Acc. # 14066 through 14072, 3255 10/1 through 10/10, 34218 5/1 through 5/5. Examples of the sheet components are 12352 20/1 through 20/20. These are roughly half the size of bells used specifically in

war. One of the examples of the latter that I have examined is the bell collected by Mohun in the 1890s, now in the NMNH, object number 174751.

[63] A blacksmith working in the Ngiri watershed area in the late nineteenth and early twentieth century traveled about 100 km to Dongo, where he trained to make welded iron bells and an elaborate form of sword. He then monopolized the manufacture of these products when he went back to his home. Mumbanza mwa Bawele, "Les Forgerons de la Ngiri."

[64] The evidence for this cultural phenomenon comes from two sets of primary sources, material and lexical. Examples of luxury products such as these are unusually well represented in museum and private collections of nineteenth century ironwork from central Africa. And although the knives especially were made in a variety of seemingly distinctive forms, closer examination of them brings out some important underlying technical features that they share. See Kriger, "Museum Collections." Specific names were coined for these fancy knives in contiguous western Bantu languages, so that one could call part of west Central Africa a "knife culture." What remains is an area of "hatchet culture," for similarly, the hatchets made as luxury goods to the south and southeast also showed patterns of shared technical features along with special names coined for them among contiguous language groups. For the knives, the languages are mainly in the northwestern and western Bantu groups A, B, C, H, and some D and K; for hatchets, the languages are mainly eastern Bantu groups L, M, N, S, and J, with some K and R from western Bantu. Where they did overlap was in the Kasai region, in the languages of Luba-Kasai and uRuund. Luxury hatchets were probably a recent addition to Tio and Pende societies. For the most up-to-date classification of Bantu languages see Vansina, "New linguistic evidence."

[65] Georg Schweinfurth, *The Heart of Africa: Three Years' Travels and Adventures in the Unexplored Regions of Central Africa from 1868 to 1871*, 2 Vols. (New York: Harper and Bros., 1874), Vol. II, frontispiece. For the history of the Mangbetu kingdoms, see Curtis Keim, "Precolonial Mangbetu Rule: Political and Economic Factors in Nineteenth Century Mangbetu History (Northeast Zaire)" (Ph.D. diss., Indiana University, 1979).

[66] Ibid., p. 93; Enid Schildkrout and Curtis Keim, *African Reflections. Art from Northeastern Zaire* (Seattle: University of Washington Press, 1990), p. 137.

[67] See, for example, the knife now in the Lang Collection of the AMNH, Acc. # 90.1/4140, illustrated in Schildkrout and Keim, *African Reflections,* p. 136.

[68] Keim, "Precolonial Mangbetu Rule," pp. 172–73.

[69] See Schweinfurth, *Artes Africanae,* Figs. and Table XVIII; Junker, *Travels in Africa*, Vol. III, p. 111; H. M. Stanley, *In Darkest Africa*, 2 Vols. (New York: Scribners, 1891), Vol. II, illustration facing p. 22; Czekanowski, *Forschungen*, Vol. II, pp. 38, 144, 290 and Vol. V, Plate 94 and pp. 35–36.

[70] See the example pictured in Schildkrout and Keim, *African Reflections*, p. 142. Copper was relatively rare in the area until the period of the Nile traders from the north, 1870 to 1885. Keim, "Precolonial Mangbetu Rule," pp. 230–33, 238.

[71] For comparative purposes, sec Czekanowski, *Forschungen,* vol. I, p. 140 and Vol. II, p. 366. For general maps of distribution for sickles, see Baumann, "Zur Morphologie." pp. 285, 295.

[72] The area includes speakers of Eastern Central Sudanic, Ubangian, and Bantu languages.

[73] I have surveyed over 240 examples in the collections of the Ethnology Section, MRAC, Tervuren, as well as others mentioned below.

[74] This statement is based on oral data included in museum accession information, and on lexical data. The sword *mukwale* was known in the contiguous languages of uRuund (K23), Cokwe (K11), Lwena (K14), Nkoya (L62), and Kimbundu (H21). Although this term is based on a root used to refer to knives and cutlasses in a number of other languages, this particular form with the meaning sword is attested in these languages only, suggesting a more recent sharing of it. For a mid-century observation of this sword, see László Magyar, *Reisen in süd-Afrika in den Jahren 1849 bis 1857* (Nendeln: Kraus Reprint, 1973. Reprint of 1859 edition), Vol. I, ill. 8. There is some evidence of its having been a luxury knife in Luba Kasai and Luba Hemba communities as well. See Von Wissmann, et al., *Im innern Afrikas*, p. 217; and Colle, *Les Baluba*, Vol. 1, pp. 185–86 and Vol. 2, Plate IX. In Yaka (H31), the same type of sword was called *mbele ya poko*, which combines a Yaka term for knife (*mbele*) with the modifier *poko*, a term for sword also shared by the languages uSalampasu and Mbala.

[75] Kriger, museum notes, technical data from the "Lunda" swords collected by Walter Currie, ROM Acc. # HAC 33, 35, 36, 37, 38; Hope Morgan Collection, ROM Acc. # HA 2472; collections of the Ethnology Department, MRAC Acc. # 17511, 17513, 22233, 23924, 30337; and "Cokwe" and "Lunda" examples in the Museu do Dundo, Inv. # D 548, 553, 1063, published in Marie-Louise Bastin, *Art Decoratif Tshokwe*, 2 Vols. (Lisboa: Museu do Dundo, 1961), Vol. 1, Figs. 93, 94, 95. See also Redinha, *Campanha Etnografica*, Figs. 127, 128, and p. 15.

[76] Kriger, museum notes, technical data from MRAC Acc. # 9387, 9391, 9392, 9399, 9404, 16074, 16077, 16078, 16079, 40084; MOM Acc. # 1907.5-28.151; ROM # HA 2680; and two examples published in A. Bourgeois, *Art of the Yaka and Suku* (Meudon: Chaffin, 1984), p. 34.

[77] This information was either observed by or recounted to a Belgian in the early 1890s, and is highly plausible as it corresponds well with the structural characteristics of the swords. See E. Fromont, "Souvenirs du Kwango et du Lunda St. Paul de Loanda" (1892–3), MRAC, Section Historique, R.G. 1120, p. 18.

[78] I have surveyed almost five hundred examples in the MRAC alone, the total falling into five main groups according to their standard forms.

[79] See Eva Dye, *Bolenge. The Story of Gospel Triumphs on the Congo* (Cincinnati: Foreign Christian Missionary Society, 1910), photo facing p. 48.

[80] It is not known how they got there, or when; the objects were not found in a stratigraphic context. K. R. Robinson, *Khami Ruins* (Cambridge: Cambridge University Press, 1959), Plate XXVII.

[81] Kriger, museum notes, technical data from MRAC, Acc. # 55.67.2 and MOM Acc. # 1908.Ty.976.

[82] See Dunja Hersak, *Songye Masks and Figure Sculpture* (London: Ethnographica, 1986).

[83] MOM, accession book, entry for 1908.Ty.976.

[84] See Stanley, *Through the Dark Continent*, Vol. II, p. 160; Cameron, Vol. I, p. 322; Hermann Von Wissmann and Paul Pogge, *Unter deutsche Flagge quer durch Afrika* (Berlin: Walther and Apoplant, 1890), illustrations facing pp. 90, 110, and p. 128; and Von Wissmann, et al., *Im innern Afrikas*, p. 277.

[85] This discussion centers on a general type of hatchet exemplified by ROM Acc. # HA 2447 and 2450 on the one hand and Acc. # HAC 572 on the other, Hope Morgan (Goad) Collection and Walter Currie Collection, respectively. I have also surveyed over 150 related specimens in the Ethnology Collections of the MRAC in Tervuren. An elaborate example was also collected by Frobenius and attributed by him to the Kalebwe

chiefdom centered at Kabinda. Klein, *Frobenius*, Vol. IV, ill. 282, p. 147. Blacksmiths in Kalebwe were known to have made such luxury goods in iron and copper. C. Lemaire, *Au Congo. Comment les Noirs travaillent* (Brussels: Bulens, 1895).

[86] Kriger, museum notes, technical data from examples in the MRAC. Of the total I examined, 18 had plain handles and no holes in the blade; 83 blades had holes of one kind or another and in different sorts of patterns; and 51 had holes with some form of copper inlay. All of them shared the same basic blade structure and silhouette, and similar wooden handles.

[87] Timmermans, "Les Sapo Sapo," p. 35.

[88] Ibid., pp. 51–52.

[89] The examples cited in this chapter and in Chapter 7 show that this stage could be marked by different levels of skill from place to place, but that there was a designated test of mastery of basic skills in each case.

[90] Maes, "La Métallurgie," pp. 75, 88–90.

PART III

SOCIAL AND ECONOMIC VALUES OF IRONWORKING: LOCAL HISTORIES

6

Patronage and Innovation in the Kuba Kingdom

Mishama nshaancy
Eyes [guardians] of the iron ore
Shósh ntúdy: Abol maam kwey.
Slogan of the smith: At home with the clan of the leopard [king].
(He belongs with the royals).

—Bushoong[1]

Having now surveyed some of the regional configurations of and dimensions to ironworkers' identities, social relations, markets, and career strategies in the nineteenth century, this next section focuses more narrowly on two case studies of ironworkers in a particular setting. Each one shows how the occupation offered men social mobility. This chapter examines the roles of ironworkers in the Kuba kingdom, where they were among the most well placed and highly respected of all the artisans who worked there. I trace some of the ways iron production and the making of finished iron products were organized and integrated into Kuba society—from the processing of ore and the patterns of iron distribution to the specialized work of smiths. I focus especially on the histories of some of their most important products, masterworks that were ceremonially displayed as symbols of status or title and were also used in the official business of the kingdom.

The main issue addressed in this chapter is the relation of ironworkers to political authority in a specific central African kingdom. Certainly ironworking was of central importance to the Kuba economy, though I argue here that it was the work of blacksmiths that was most highly developed and prized. While there is evidence that iron ore and smelting were indeed valuable, blacksmiths contributed substantially to regional trading networks of the kingdom, and above all, to the bureaucracy and systems of patronage within it. Smiths produced important insignia and symbols of status that marked titleholders and elites, visual forms that reflected and articulated the complex layers of stratification in Kuba society. Hence smiths were integral to the workings of the kingdom's administration, though it is unlikely that they were ever under strict royal control. Instead, the exceptionally high level of training and innovation that is consistently evident in their products suggest that smiths controlled their work and competed with one another for favor. Contents of the kingdom's treasury reveal that the most ambitious and brilliant masters were rewarded by royal recognition and patronage.

ARTISANS IN THE KUBA KINGDOM

Nowhere in central Africa did artisanal production of commodities reach higher levels of achievement, in quality and diversity, than in the Kuba kingdom. At the end of the nineteenth century, wealthy men from the kingdom who traveled outside it were instantly recognized by their proud carriage and above all, by their elaborate, impressive dress and equipage.[2] Although rare luxury goods from foreign lands and overseas were keenly appreciated at the Kuba court, so, too, were the best manufactures made by craftsworkers within the kingdom. Artisanal work created an array of value-added goods that rewarded wealth and service, thereby helping to generate prosperity. And outside the kingdom, the apparent splendor and desirability of Kuba products offered its traders special leverage for turning profits in regional commerce.

Situated near the center of the African continent, along the southern edge of the rainforests and at the confluence of the Kasai and Sankuru Rivers, the Kuba Kingdom had been only indirectly linked with overseas trade networks before the nineteenth century. No archaeological work has been carried out there to date, and so what is known about the kingdom's early trade relations comes from brief references in oral traditions and testimonies, from linguistic data, and from inferences that can be drawn from art objects and other material evidence that were preserved in the king's treasury. Cowries, one of the currencies used in international trade, appear on statuary associated with eighteenth-century kings, for example, suggesting that commercial links with the west coast were becoming better established and perhaps more frequent during that time.

The kingdom was very well placed and prepared to profit handsomely from foreign trade during the nineteenth century. As Imbangala and Ovimbundu merchants pressed the frontiers of the Atlantic trading zone steadily eastward into central Africa, Kuba became a middleman source of ivory for them. By at least the 1830s, Luso-African caravans were arriving at its southern borders where their cowries and trade beads purchased tusks brought by Kuba merchants from the upper Lukenie, north of the kingdom.

The ivory wealth of the inner Zaire basin thus became a valuable resource area for external trade, and an area which, although formally outside the Kuba king's domain, contributed more and more to its already complex and dynamic economy.[3]

It was an economy that had long been fed by production for regional trade networks, and while overseas trade gained in prominence it did not displace these earlier commercial relationships. Trade between the kingdom and its neighbors continued to operate and fuel economic growth. Cast copper cross-shaped ingots from Shaba and copper wire were imported into the kingdom from the southeast, some of it amassed in the royal treasury or used by artisanal workshops in the capital to enrich and embellish court paraphernalia such as the dynastic drums.[4] Copper was also exported to the north and northwest, for ivory and for the lower cost manufactures of their less prosperous neighbors. Hence her middleman position in nineteenth-century trade owed a great deal to a foundation built upon earlier, regional precedents.

The success of the kingdom's middleman role cannot be sufficiently explained by an ecological model of trade. Certainly the environmental diversity and natural resources present had a lot to do with the strength of the local economy, especially in the agricultural sector.[5] However, it was human labor and human capital that transformed her natural endowments into economic advantage. Much of what was traded regionally consisted of processed and manufactured goods, that is, they were not simply gathered or harvested raw materials. Salt, for example, was produced within the kingdom according to at least two different methods, one of them considered to yield a better quality product. Other kinds of salt were imported from several districts in Shaba, and salt was also exported from Kuba northward in exchange for goods such as ivory and probably iron. This is only one of a number of examples that could be given showing that domestic and regional trade was not based simply on the presence or scarcity of certain kinds of natural resources. Goods of various types were made domestically *and* were imported and exported as well. Examples of manufactures include objects carved of wood, different types of plain and fancy cloth, pottery, and ironwares—all of these made inside and outside the kingdom and exchanged regionally.[6] Production specializations, distribution networks, and the tastes and preferences of consumers were all sufficiently complex by the nineteenth century to maintain a robust regional trade, a trade in which the specialized skills of artisans clearly played a major role.

Not only were artisans important to the Kuba economy, their products were important to social life and to the political workings of the kingdom. Manufactured products served more than immediate, practical functions. Many of them were worn, used, or displayed to indicate one's social rank or current political office, for example. The elaborate systems of social relationships, hierarchies, and bureaucratic organization that had developed over time within the kingdom required equally elaborate systems of visual identifications, references, and verifications. Indeed, it was the stunning array of finely crafted manufactures and art objects spawned by these visual systems that made the Kuba kingdom so famous in Europe and North America by the beginning of this century.[7] Oral societies, much more so than societies with written forms of verification and documentation, tend to rely heavily on visual imagery and objects to perform a host of communicative tasks, and the Kuba kingdom was a prime example that shows the degree to which such needs could be elaborated and satisfied. Consumer demand for certain manufactures in the kingdom must therefore be understood not in strictly economic terms, but above all in terms of the social and political values they represented and upheld. Thus the work of smelters, important though it was, had been eclipsed in the Kuba kingdom by the even more remarkable skills of her blacksmiths.

PRODUCTION AND DISTRIBUTION OF IRON

There is no geological survey that could provide an overview of the extent and location of iron ore deposits in the Kuba kingdom. Shallow deposits of laterite seem to have been not uncommon, the ores exposed along riverbeds and gathered during the dry season for smelting. Such was done in the northeast portion of the kingdom settled by Ngongo subgroups, and it was also described for the southern part of the core area, around Ibaanc.[8] In at least two locales there was mining of more concentrated deposits, though only one of them was formally incorporated into the kingdom proper. That was the mine at Mboong Bushepy, just inside the southwestern frontier of the kingdom. Kuba rulers periodically exacted tribute from villagers there in the form of spherical bar iron units, assessed at five to ten units per adult man.[9] The mines around Kabuluanda, near the southeastern border, yielded ores that were rich and plentiful, but how much was ever sent to the capital as tribute remains unclear.[10]

Notwithstanding the high value that seems to have been placed on iron and iron products in the Kuba kingdom, specific references to ore deposits and tribute in iron were usually left out of oral traditions. The examples of resource areas described above are only those that were best known or were observed by outsiders. As such, they are merely an indication of what must have been a variety of ores and deposits that were available for exploitation within the kingdom and also along its periphery. Even with such incomplete information, though, it is worth revisiting the question of whether or not it was considered vital for political leaders to directly control iron resource areas.

The Kuba case suggests that it was not. Instead, the *modus operandi* of Kuba officials appears to have been consistent with that of most merchants in the region, who formed trade or tribute agreements with ironworkers in order to take a percentage of their production rather than imposing direct control over the ore. The only instances known so far that might be said to contradict this suggestion were the attempts to control the Kabuluanda mines. Oral traditions claim that the southward expansion of the kingdom during the seventeenth century came to a halt when the territory around the Kabuluanda mines was taken, and that it was successfully defended from Luba seizure by forces of King Kot aNce in the eighteenth century. During the period when the mines were within the kingdom's borders, access to the ore was administered by the Pyaang aTyeen, so the Bushoong ruling elite had never in fact established their own direct control of the iron resource itself. By the beginning of the nineteenth century, however, Kabuluanda had changed hands and was under some form of Luba administration, though this loss was not mentioned in Kuba oral traditions.[11] Whether the miners and smelters who had worked the ore deposits remained there or were completely replaced by others is not clear, nor is it certain that Kuba ironworkers never again managed to receive supplies of iron from the new overlords. The pattern that is suggested here is that resources such as iron ore and iron were sought most often through negotiation, not through territorial conquest. Battles for direct control were rare probably because they threatened the autonomy and cooperation of ironworkers, and therefore also the continuity of iron production. Flexibility in seeking access to a portion of the product through tribute or trade appears to have been the more prudent option in the long run.

However, it is not yet possible to estimate what the volume of iron production within the kingdom might have been. No surveys of smelting evidence have been conducted, no smelts were observed first hand, and there is only one description of a furnace. In 1908, at the town of Misumba in northeast Kuba, a master smith recounted the way smelting had been done. He claimed that in times past there were furnaces working in each village, and the furnace that he described was relatively small in scale. It was a bowl of about 50 cm diameter and depth, with air forced through two pairs of bellows set inside a single tuyère. Termite clay was used in some way, but it is not clear from the description and diagram whether it was used to make the tuyère or to insulate the furnace bowl. Ore came from river bed deposits, and the actual smelts reputedly lasted two days and two nights.[12] Whether there were other furnace designs used by smelters in this or any other part of the kingdom is not yet known.[13]

This same smith also told the Ngongo version of the origin of iron:

> Ironworking was taught to men by certain spirits who appeared to them in a dream, saying: "What, you are a strong people and you travel unarmed?" The spirits showed the men a certain river named Mosanja, and ordered them to dig up the soil; then the men were told to gather up some termite

> clay, an amount about the size of a man's head. The spirits then taught the men how to construct a furnace with the termite clay, and how to smelt the ore that they got from the riverbed soil.[14]

The supposedly divine origins of iron metallurgy in this account suggest that smelting in this part of the kingdom was of some antiquity, ascribed to a time before historical memory. Linguistic analyses also point to a knowledge of iron metallurgy already shared among some proto-Kuba groups as they settled the lands south of the Sankuru.[15] However, details in the account which correspond to descriptions of more recent practice, such as the use of termite clay, are anachronisms added to perhaps an older tradition.

It is also possible that the account was embellished with references to divine intervention in order to parallel traditions maintained at the court of the Kuba king. The Mwaaddy, eldest living son of the king and keeper of royal traditions, told the Bushoong version of the origin of iron:

> One day Woot found a large stone that had been left behind by Mboom [God]. "What is this?" he asked. The people answered, "It is Mboom's shit." So Woot had it carried to the village and honored. During the night, Mboom appeared to Woot in a dream, and said to him: "You have acted wisely, honoring all that comes from me, including my shit. In return, I will teach you how to use it." And so Mboom showed Woot how to smelt iron.[16]

Taken at face value, this story could be interpreted as a glorification of Bushoong paramountcy, and so it must be viewed with suspicion. For the same reason, other claims that either the Bushoong or other proto-Kuba groups introduced improvements in ironworking to peoples in their new homeland are also suspect.[17] In short, much of the oral evidence referring to iron in the earliest periods of Kuba history is not reliable except as a general indicator of the perceived antiquity of ironworking and the high value placed on it by the composers and amenders of the oral traditions. Given the lack of archaeological work in the kingdom, and the intermarriage and assimilations that took place between the ethnic subgroups who inhabited it, the origins and chronologies of early Kuba iron production remain unclear.

Oral traditions explaining the creation of kingship portrayed the political struggle between chiefdoms as a competition between various leaders on the plain of Iyool. There, the Bushoong leader prevailed in a test against his main rival. In variants of the story, each of the leaders was to make a metal object, a copper plate, iron axe, hammer, or anvil, and throw it into a lake or river. The one whose object floated would win. In each case, the trick was to make the object of wood and cover it with sheet metal; sometimes the Bushoong leader made the metal-covered object himself, other times it was stolen from his rival by a woman.[18] To be sure, this episode was not designed

Fig. 6-1: Kuba king figure, Mbop Pelyeeng aNce. This king was said to have been a blacksmith, and selected an anvil as the sign of his reign; however, since the sign-object (shown at lower center, in front of the figure) is damaged, it is not clear whether or not it is indeed an anvil. Like other carved Kuba king figures, he holds the royal sword in his left hand. Wood, 55 cm high. Copyright © British Museum, Negative number MM033101. Museum of Mankind, London, 1909.5-13.1. Collected from the royal treasury in Nsheng by Emil Torday, British Museum Congo Expedition 1907–9. [Photograph by the British Museum Photographic Services]

to tell about ironworking in early Kuba history, but it does make one very important and valid point. It was finish work in iron, not so much smelting, that Kuba ironworkers excelled at and were known for, and so it is fitting that in the story of kingship the early central Kuba groups and the Kete autochthons performed clever feats of blacksmithing.

Equipment used in the Kuba finishing forge presents some convincing evidence of technological improvements that probably came originally from southern Mongo workshops. Blacksmithing tools collected at the beginning of this century convey how important the smithy was in Kuba society, for it is evident that there had been concerted efforts to upgrade ironworking toolkits. At least two different types of iron anvils gave smiths a range of options for producing finely crafted and detailed products. The heavy sledge hammer/anvil, *iloon*, when used as an anvil, was inserted into a wood base, which secured it well and which itself could also be used as a surface for hammering and drawing out the basic form of the product. This particular method for using the sledge/anvil along with a wood base was probably developed in the vicinity of Malebo Pool, perhaps as long as several centuries ago, and then was adopted by blacksmiths working in southern Mongo communities. Another anvil in the Kuba blacksmith's toolkit was much smaller, designed so as to allow the smith to work at an even finer level of detail. The precision so evident in the structures of certain Kuba masterworks was made possible by these anvils, along with a special hammer. The finishing hammer, *nsháák*, already discussed in Chapter 5, was ingeniously designed for multipurpose drawing out and detailing work. It could be gripped at one end and struck longitudinally, or gripped in the middle with blows struck at a perpendicular angle. Each of its two working ends was shaped and edged differently, thus giving the smith greater flexibility in performing specific types of hammering tasks. This particular tool, along with its lexical term, differs significantly from the finishing hammer used by smiths in the former Tio kingdom and environs, and so most likely was invented and developed over the past several centuries by southern Mongo smiths.[19]

The Bushoong ruling elite claimed blacksmithing as their forté, and stressed its importance at the capital. Before the twentieth century, all men of the royal lineage were supposedly trained in smithing to some degree, and at least one king, Mbop Pelyeeng, was said to have been a smith himself. The emblem he chose to signify his reign was an anvil, as is indicated by the carved commemorative statue of him that was kept in the royal compound (Fig. 6-1).[20] Other smiths whose names were remembered were Myeel and Pyeekol, who reputedly lived and worked at the court in the seventeenth century. Some recounters of oral traditions claimed that Myeel, a member of the royal house, was considered a virtuoso smith but was so strict that he was not chosen to be king. Others stated that his mother invented smithing.[21]

The work of smelters, blacksmiths, and other artisans who resided outside the court and in the provinces was monitored (but not managed) by government officials. Many of the free male inhabitants of Nsheng, the capital, were titleholders whose job it was to keep informed about the production of valuable goods and resources so that additional tribute might be collected as needed for supplying and enriching the king and his court. Many of these titleholders were themselves smiths, as would have to have been the case for the *tancoon* and his deputy, the *cikl tancoon*, who were the officials in charge of iron ore.[22] Not only did officials like these keep an eye on workshops and their output, they also maintained order and settled disputes that arose in markets and other commercial settings.[23] Knowledge of the craft, especially how to assess quality and value of the iron product, would have been an essential aspect of their authority.

Although certain ironworkers were subsidized and others monitored by the Kuba court, it would be an exaggeration to say that ironworking was consistently and strictly controlled by the central administration. Most likely, as different kings formulated their own political and economic policies, so, too, they would recalibrate their tribute and trade relations. Iron production certainly would have been more crucial to national interests in times of war. There was the village of Paam, for example, which apparently was reduced to matoon (unfree) status by King Mboong aLeeng for not reporting their iron output.[24] That this was said to have taken place during the seventeenth century, a time of territorial expansion, may be key to understanding the king's reaction. Ironworking then may have been very closely monitored, given the importance of iron to weapons manufacture and supply, and thus would explain the harsh punishment reputedly endured by the entire village.

But if there had been a consistent policy of controlling iron production in the kingdom, one would find remnants of a coherent, centralized pattern of workshop locations and organization. Instead, the pattern that one does see is more variable and complex, and suggestive of competing centers of production. Certainly there was important iron manufacturing going on in the capital, but there were also cases of impressive manufacture and innovation outside the core area of the kingdom. There were Kete blacksmiths, for example, in the southern Kuba town of Kakenge who were famous for their skill, and Pyaang smiths were considered by many to have been the best of all.[25] Moreover, the iron flowing into the capital as tribute was not just raw iron to supply smithies there. Some of it was bar iron, some was already worked into finished, standardized products such as razors and hoes, and some was in the form of more unusual products as well. Prestige goods, such as *ikul* knives and certain types of war knife, were made in provincial workshops and examples of these were selected and sent as tribute to be distributed by the king and his officials. *Ikul* was a class of knife that was challenging to make, and it was considered an essential feature of the equipage of male titleholders in the kingdom. Surplus quantities of them were made by blacksmiths in Pyaang and Ngeende communities, and to a lesser degree Coofa and Kete as well.[26] What the varied patterns of iron production and distribution suggest is that blacksmiths were monitored and encouraged by government officials, but they remained in control of their workplaces.

As the kingdom became more prosperous over the eighteenth and nineteenth centuries, consumer demand for prestige goods grew and craft manufacture became ever more specialized. That and the general trend toward greater population density in Kuba lands must have placed strains on the ability of smelters to produce sufficient iron supplies for both utilitarian and luxury needs. Hence iron in all sorts of forms probably came increasingly from outside the kingdom, though again the evidence is sparse and spotty. Shoowa and Kel communities in the northwestern corner of the kingdom imported iron, though the ultimate source of the iron was not specified.[27] Bar iron and blade currencies circulated in areas all around the kingdom in the nineteenth century, some of it probably finding its way to Kuba forges. It may well be that the Sakata ironworkers who had moved their workshops to the upper Lukenie River by the late nineteenth century did so in order to supply iron to Kuba blacksmiths and other consumers. Other external sources of iron could have been trading partners to the south and southwest. One observer in the early twentieth century noted that Kuba blacksmiths were able to purchase their refined iron in markets, giving as an example the town of Ibaanc, which at the time was the entrepôt for goods coming in from Lulua, Shaba, Bihé, and from as far away as the west coast. In the town market, bar iron was available, hammered into units weighing two to two and one-half kg each.[28] Apparently there was not enough domestically produced bloomery iron to support the manufacture of iron goods in Kuba, at least not during the nineteenth century.

Whether Kuba kings considered it an option to try and retake the Kabuluanda iron mines is unknown. It would seem, though, that Kuba officials and blacksmiths did find other ways to compensate for the inadequate supply of iron in the kingdom. Iron was generally kept out of its own internal currency markets, the major precolonial currencies having been raffia cloths, cowries, and then in the late nineteenth century, also beads and copper ingots.[29] And in the category of prestige goods, which would have placed heavy demands on iron supplies, there was a trend toward substitution of other materials for iron. *Ikul* knives were in some cases made out of imported copper. In still other cases, elaborately carved replicas of these knives and also two types of war swords were made entirely out of wood.[30] Scarcity and hence conservation of iron seem to be the most reasonable explanations for why woodcarvers would have fashioned products so precisely and cunningly in imitation of valuable metal ones.

IRON INSIGNIA AND SYMBOLS OF STATUS

The more elaborate products of Kuba finishing forges were distinctive, set apart from the work of smithies in surrounding areas and throughout the region. As highly valued prestige objects, they and other domestic manufactures can be read as material manifestations of the ways ambition, wealth, and social position together fueled what has been called the "Kuba miracle," a society noticeably more sophisticated and refined than its neighbors.[31] Production and consumption of manufactures were essential to its economic well-being and its collective identity. An American missionary, new to the area, recorded his early impressions in 1904:

> As to the Bakubas' praiseworthy industry, I can say that I have rarely found a Bakuba idle. But a few months ago I would have asked: 'What, pray, do they find to do?' It may surprise you to learn that they have plantations of corn miles in extent; that they raise peanuts and manioc and vegetables which require much work; that they have chickens and ducks and goats; and that they do much hunting and fishing. But greater still was my surprise to see a well-organized court in every village, with judge, lawyers and jury. I did not expect to see blacksmith shops with smelting furnaces for crude ore, with anvils and bellows and hammers; nor the unique handlooms turning out various kinds of really exceptional cloth; nor men weaving beautiful mats with variegated patterns, while others prepared tobacco or made salt from the leaves of a swamp bush, or built capacious dwellings, or made paint and dyes, razors, weapons of war, cooking utensils, hats, sundry ornaments; indeed, scores of different things.[32]

Fancy metalwares were prominent among the many types of luxury goods made within the kingdom. From the metal hatpin, *ndoong*, which the Bushoong cited to claim they were more civilized than other peoples, to the double bell, *mikop*, an insignia of the king in his role as Bushoong chief,[33] certain products of the smithy were made expressly to serve an array of social and political purposes. They did not constitute a fixed system of signs of office, but together with other handcrafted objects and rare materials they made visual reference to current titled positions and social relationships, and thereby amplified over time the ideology of the king and his supporters.

As surviving historical documents produced within the kingdom, products manufactured for the luxury market are worthy of attention in and of themselves. They can also offer some informative glimpses into the history of Kuba ironworking, though not all in the same way. Kuba luxury goods fall into two general groups, each group having a particular kind of social and political significance: objects that were worn or carried by men as their own personal insignia of office or symbols of status; and the objects kept in the king's treasury that were carried by men as verifications that they represented the king or one of his officials. The latter will be discussed in the next section, though it will soon become evident that it is not always entirely clear into which group an object ought to be placed. All of them, their strong visual impact and their obviously significant uses, must have lent some degree of privileged status to their blacksmith makers.

These documents thus have much to say about social stratification and the political deployment of elaborate luxury products in the kingdom. At the same time, they also convey to us invaluable details about the development of skills, design innovations, and other changes in the work of blacksmiths over time. But before I present my analyses of them, I must insert a note of caution. Certain metal products, even some rare ones, are not difficult to identify, for they were widely known and well established ceremonial objects available to prominent men in the kingdom. Others, however, present intractable problems. Ceremonial objects that show some clear evidence of Kuba workmanship but are poorly represented in museum collections are difficult to evaluate as to their purpose and degree of rarity. One such example is the hoe or adze now in the Hampton University Museum, with its blade elaborately decorated by incised patterning much like that found on Kuba textiles.[34] Equally problematic are objects that were reported to have been insignia of office, such as certain axes or adzes, but which, again, are not included in museum collections and so cannot be further identified or analyzed.[35] What I have chosen to do here is to focus only on the most well-known ceremonial objects and examine them closely, noting how they were made and what they were called, so that they first can be arranged into a relative chronology. And in so doing, other general trends in the history of Kuba blacksmithing will also come to light, showing once again that ironworkers, and their techniques and products, were not confined within family, residence, or language groups. For these special products of Kuba smithies exhibit features that can be traced back to diverse origins in southern Mongo, Luba, and other neighboring societies.

The Kuba knife *par excellence*, the one whose display identified a broad class of high status and respectable males, was the personal weapon known as *ikul* (Fig. 6-2).[36] How much it was used as a tool or weapon is debatable, for most of the examples now in museums show signs of wear only on the handles, and not on the iron blades. Although there have been no attempts to try and date any of the blades, it would not be surprising if some were very old, inherited from forebears and remounted into new handles.[37] The *ikul* seems to have been used most of all, at least in the nineteenth and twentieth centuries, as an essential component of costume for free, well-placed adult males. It was not large, the knife and handle together averaging only about thirty-five cm in length, and it was worn at the waist, tucked into the man's belt. The Rev. William Sheppard, who lived in and near the kingdom from 1890 to 1910, recalled several times how important it seemed to have been

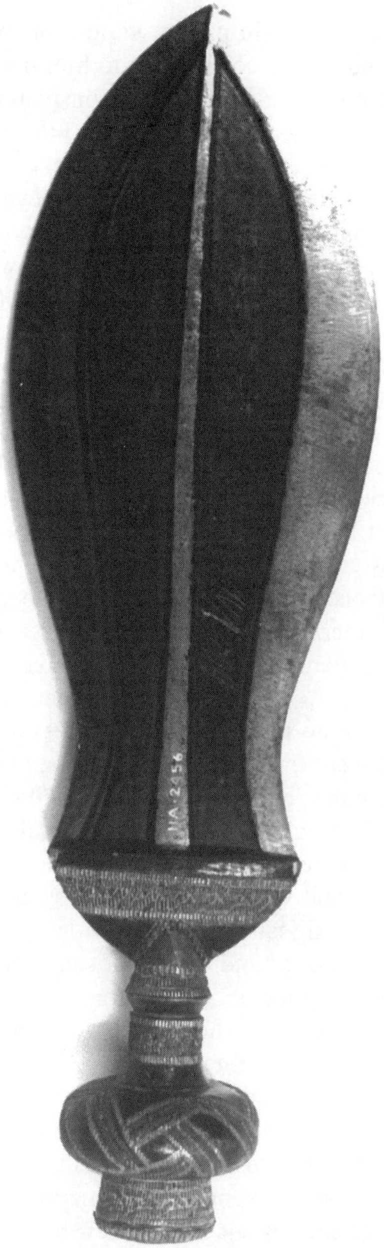

Fig. 6-2: *Ikul* knife, part of the equipage of aristocratic adult males in the Kuba kingdom. Iron blade, blackened and polished, with metal encrusted handle. Blade, 26 cm long. Courtesy of The Royal Ontario Museum, Toronto, Canada. HA 2456. Probably collected by William Sheppard, then given to T. Hope Morgan before 1905. [Photograph by the author]

for a free man not to be seen in public without one of these knives as his personal ornament, and how crucial it was to him not to lose it.[38]

At first glance, it could be judged as inconspicuous and quite plain in appearance. Most frequently, the blade of the knife in silhouette was like a rather broad, gently curving leaf, its surface blackened and selectively polished to highlight the midrib and one of the two beveled edges. There were also several other variants within this general knife type that had been developed by inventive smiths over time. One variation was relatively rare, the surface pattern consisting of four or five wide, parallel blood grooves covering the right half of each face of the blade. Another surface design was an all-over chiseled pattern resembling the roughly textured back of a tortoise. Such innovations within this special category of knife suggest a considerable age for the original prototype, and that it was probably the plain leaf form that had been invented first, since the variations all had to do with additional, more elaborate surface patterns, not changes in the basic structure of the blade.

Ikul knives ranged in quality from the most elegant or fancy to simpler ones for everyday use, the wealthier and more prominent men having had several for different occasions.[39] Certainly what distinguished the higher quality knives from others were the iron blades themselves, which differed by how skillfully they were made, though even greater differences could be seen in the workmanship of the handles. Simpler blades of lower quality were usually mounted onto plain wooden handles, which, although artfully designed and carved, were finished only with a coating of oil. Fancier knives had the same basic handles, but they were decorated with delicate patterns of checkerboards and interlaces, made from fine metal strips of copper, brass, or zinc that were hammered and chiseled into the wood. The result was a much more impressive and valuable product, the fine patterning of the embellishments reflecting a signature aesthetic of Kuba craftsmanship. Similar motifs, elegantly proportioned and executed, can be seen in Kuba cloth embroidery and relief carving on wood. Such a range of elaboration in the workmanship of these knives suggests how finely tuned Kuba social hierarchies could be, even within the male elite.

Upon closer examination, however, it is not the skilled metalwork decorating the handles of fancy *ikul* knives that is most admirable about them. Less immediately accessible to the eye, especially to one unpracticed in smithing, is the cross-sectional geometry of an iron blade. Turning our attention to this feature, we see revealed before us the blacksmith's design process and the forging skills necessary in manufacture. In the case of the *ikul* knife, it is the exceptional structure of the blade, the way the blade was designed and modeled, that serves as evidence that there were singular intellectual and manual abilities developed among Kuba blacksmiths. The formation of the central, offset midrib and the way the iron was pulled off from it on either side, stepping the metal down and then up to the beveled edge,

Fig. 6-3: Public display of *iloon* sword in the Kuba capital of Nsheng. Copyright © British Museum, Negative number MM003625. [Photograph by Emil Torday, British Museum Congo Expedition 1907–9]

were very challenging and difficult tasks for even a highly skilled smith.[40] It produced a structure that, as can be seen in the subtle modeling of its forms, went far beyond the practical necessity of strength. One could go so far as to say it implied the existence of a knowledgeable, discerning clientele, and producing such a knife as this may have served as proof that an apprentice was ready to establish his own forge.

The elegance and ambition so evident in the *ikul* knife's design, and the steady control demanded in its manufacture, set Kuba blacksmiths apart from all others in the region. Whoever the inventor of this remarkable design was, he was a superb master, probably from among one of the core proto-Kuba groups, and it would not be unreasonable to suggest that he lived just before or during the seventeenth century.[41] That such an intricately forged product was then so carefully replicated in numerous workshops over time testifies to unusually high collective standards of excellence and skill levels maintained by the kingdom's blacksmiths. In other words, the respect they garnered was visibly well earned.

In contrast to *ikul*, the Kuba war sword, *iloon*, bears evidence of foreign origin (Fig. 6-3). Usually referred to as a ceremonial or parade sword, most examples had grooved and incised patterns on the iron blades, with wooden handles decorated by sheet metal, shells, and beads. When the entire sword was made of wood, these specific features and materials were convincingly reproduced by the woodcarvers, indicating that they were imitations of the metal prototype.[42] Two lines of evidence suggest that *iloon* swords entered the Kuba blacksmithing repertoire from a southeastern or eastern source, either from the Luba or elsewhere, through Luba intermediaries. The conventions of surface patterning exhibited on the blades are the opposite of those used on the Kuba *ikul* knife, and are like those used by blacksmiths who worked in the upper Zaire, Lualaba, and Lomami River basins.[43] The lexical term *iloon* appears to have been a loan word introduced into Bushoong from Luba Shaba or Songye neighbors.[44] The visual and lexical data together indicate that the *iloon* sword was originally invented outside the Kuba kingdom.

How and when it came to be produced in Kuba workshops and why it was no longer produced in workshops outside the kingdom becomes, then, a matter of historical interest. This particular type of sword was a consistent feature of the Kuba king in carved commemorative statuary, where he was shown holding the sword in his left hand as he sat cross-legged and serene on the royal dais. It may well have been intended as a mnemonic to refer to the glorious period of Kuba expansion in the seventeenth century, and the military strength and successes over Luba-speaking rivals in the south and southeast. Iron *iloon* swords were massive enough to have been used as weapons in the past, handle and blade together usually measuring from fifty to sixty-five cm long. But they were not necessarily easy to replicate. Indeed, the smiths who were expert in making these particular swords must have been either captured in battle or invited or coerced to come and work at the Kuba court, perhaps as early as the seventeenth century, but more likely in the eighteenth. From that time onward, the best *iloon* swords were associated with the king, who customarily held one of them in his investiture ceremony, while he recited the list of Kuba kings and capitals.[45] High ranking officials were also permitted to carry and display them, and members of the royal entourage danced with these swords in hand at public ceremonies. In the nineteenth century, they were made only in the Kuba kingdom, by Pyaang blacksmiths and by smiths in the core area and at the capital.[46]

Two other types of knife used for ceremony and display in Kuba were not so closely identified with the kingdom in the nineteenth century. Moreover, each type shows more variability in form and in the special social and political conditions of use than *ikul* or *iloon*, suggesting that each had been incorporated more recently into Kuba blacksmiths' repertoire. Together they reiterate the multiple historical origins of blacksmiths' products, markets, and skills. One of these was a knife or sword referred to as *ngodipy*, displayed by specialist healers and sometimes required as part of the payment owed to judicial authorities hearing cases of capital crimes.[47] Some examples of this sword were indistinguishable from products of Kete, Sala Mpasu, and Lulua smithies.[48] Two other specimens, collected by Torday in the Kuba capital, were morphologically related to the others but also show signature features of Kuba manufacture.[49] One was a masterwork of smithing, the iron blade decorated in an allover pattern of concentric half circles of copper inlay, the handle carved and decorated with sheet metal as were other typically Kuba products. It is likely that this knife was made in a royal workshop.

Another type of knife, similar in scale to the *ikul* but with an unremarkable structure, is only sometimes clearly identifiable as a Kuba product.[50] Usually referred to as a dance knife, it was associated with the male elites and leaders not only of the Kuba kingdom but also of neighboring Leele, Dzing, and Songo Meno communities. It may well have been invented elsewhere and imported into the kingdom at one time, whereupon highly skilled court smiths at Nsheng could easily have replicated it.[51] Many examples were made of wood, some to be later decorated with cowries or other valuable materials. Known as *ikul imbaang* in Bushoong, these may have been the knives that were traded outside the kingdom for Leele cloth and carvings, and perhaps also for ivory from the north.

In certain cases, the *ikul imbaang* blades were made of iron with copper medallions inset at the center, while in other cases the entire blade was of copper. An unusually complex specimen, made of copper but with the offset midrib typical of the older *ikul* knife, was probably also a product of court workshops.[52] Copperworking was done mainly at Nsheng, the capital, and required special training in the handling of that softer metal. At the same time, new techniques for mixing metals and for casting had to be mastered by blacksmiths there. Whether such advancements came about by experiment or through training with immigrant smiths or both is unclear, but it was not a simple process. The copper medallions on *ikul imbaang* knives resembled the openwork done on Songye hatchets and knives, but the specific techniques used in each case were completely different. Moreover, the casting methods used for making bracelets in Kuba smithies appear to have had historical links to copperworking west of the kingdom,[53] while the copper itself was, at least in the nineteenth century, imported from the southeast.

Still other Kuba metal products, very special ones, call attention to the social and political importance of blacksmiths' innovations. At the center of the brilliant and visually stunning Kuba universe was the king, and certain rare products and materials were reserved for him. An example was the large knife associated with the king from at least the late nineteenth century onward. Called *mbombaam*, it was reputedly among the king's own insignia, for him to display on the most important public occasions. The example that was acquired by Torday shows it to have been a knife quite different morphologically from the others that have been discussed above. It was a single-edged blade, unlike all the others; it was larger in scale than the others; and it was unusually ornamented with nine parallel blood grooves along one face, the other face simply being the negative. The wooden handle and the metalworking patterns worked into it were, however, characteristic of Kuba workshops.[54] It is probable that the blade was an innovation designed especially as a singular object, ideally suited therefore to grace the king. Interestingly, though, there were at least two replicas of this knife made in wood that were observed in Pyaang communities, though no information was given as to how they were used or by whom.[55]

The relative chronology suggested by my analyses of these complex metalwares reveals a pattern of endogenous and exogenous innovation at the blacksmith's forge, encouraged in part by royal patronage. The product type most likely to have been the oldest, the *ikul* knife, displayed a particular kind of virtuosity which demanded expert control and precision in drawing out and modeling a mass of iron metal. Its design was a masterpiece of understatement, fully appreciated mainly by connoisseurs of metalworking and other smiths. Quite a different set of skills and techniques were applied by smiths in forming the *iloon* sword, a product of foreign origin that was large in scale and more overtly spectacular, embellished with engraved, punched, and stamped patterns. The *ikul*, exhibiting design conventions developed by smiths of southern Mongo groups who settled south of the Sankuru, and the *iloon*, exhibiting quite different conventions from Luba-speaking areas or beyond, are thus three-dimensional historical documents. They represent in material form the major cultural groups whose meeting and interminglings lay at the foundation of this kingdom.

During the eighteenth and nineteenth centuries, Kuba blacksmiths appear to have produced several additional types of knives or swords to satisfy a growing need for new prestige goods and royal insignia. Two examples were based on products that were invented outside the kingdom. Swords called *ngodipy* in Bushoong came originally out of smithies to the south, in Kete, Sala Mpasu, and Luba Kasai communities. Since specialized skills were necessary for making them, it is likely that Kuba blacksmiths retrained in smithies outside the kingdom, or foreign smiths were encouraged to immigrate. The knife *ikul imbaang* probably originated in smithies west and southwest of Kuba, though it was a type of blade that skilled blacksmiths could replicate from an imported one. Kuba smiths developed the product further by working it in copper—either the entire blade, or just the medallions sometimes set in its center. As the kingdom grew, then, techniques and skills appear to have been drawn into the Kuba capital from a number of metalworking traditions in the region. And finally, the last example of innovation was the great knife of the king in the nineteenth century, *mbombaam*. Larger in scale and structurally distinct from all other knives and swords known, it was probably a relatively recent invention created by smiths at court for display on royal occasions. These knives and swords represent the period when the kingdom was at its height, when wealth and political power became increasingly focused around the court. Together, all of these metalwares demonstrate the complexities of blacksmithing technologies over time, and the effects these technologies had on Kuba society. Innovations came from both internal and external sources. Blacksmiths and their government patrons sought out new techniques, skills, tools, and ideas from their immediate and more distant neighbors and worked to improve on them, making new products they called their own.

MYEEL, THE GREAT SMITH

Collective pride in the kingdom's civilization was officially expressed at times through the presentation of special gifts to foreigners. Some of these gifts were the most extraordinary feats of metalworking performed by master blacksmiths of the past. Selected from smithies in the capital and probably also from workshops elsewhere in the kingdom,[56] virtuoso works of art in metal were kept in the treasury along with other riches accumulated through tribute and trade. They embodied the wealth of the kingdom, some of the glories of its history, and the taste, wit, and acumen of its foremost master blacksmiths. Amassed incrementally over time, this was a collection of masterworks made by many different smiths, their accomplishments telescoped into the products of one great forge and one great master, Myeel.[57]

One-of-a-kind masterpieces were not just glorifications of the king and the kingdom, or of the blacksmith makers, but served more practical purposes as well. Singular objects that were known to belong to the treasury were used in all sorts of ways on ceremonial occasions and on official business. Some were carried, for example, to identify royal messengers relaying information to and from the king or his subordinates, while others verified the legitimacy of government tribute collectors as they traveled from village to village in outlying provinces of the kingdom.[58] Such visual confirmations, in the form of technically complex objects that could not be replicated, were essential tools for administering government effectively in an oral society.

One group of very unusual objects was gone from the treasury by the early years of the twentieth century, but they were still remembered as famous products that had been made by the great Myeel. These were sculptures forged out of iron, including human and animal figurines, a boat complete with crew, and a miniature house.[59] How many such objects there were originally is not known, but three of them ended up in European collections. One, a hunting dog, is in private hands and has not been fully described or analyzed.[60] The other two, human figures about twenty cm high, exhibit a number of interesting features. They were indeed forged, the separate parts of the figures assembled and joined together by high heat welding,[61] techniques also associated with the making of iron hammers and anvils. One appears to have been damaged, suggesting that the other was made to replace it. To my knowledge, scholars have not noted the intriguing pose of the figures, specifically the unusual placement and angles of the arms and hands repeated in each example. Given the importance of posture already demonstrated in the well-known carved wooden king figures, what seem to be certain hand gestures being articulated by the iron figures could represent an iconographic message or cue that may no longer be understood. Only commented on in passing was the pierced opening at the top of each figure's head, a feature that would have allowed the object to be worn suspended at the waist or from the neck. It is probable, therefore, that these figures were among those objects used in official displays and perhaps also worn on the king's person or by certain of his emissaries.

According to the oral traditions, Myeel was also credited with the skills for combining different metals into the same object. One such object, formerly a part of the treasury at Nsheng, was an exquisitely made anvil, obviously intended only for ceremonial use.[62] Made of several constituent metal sections joined together, it was modeled after a small finishing anvil, the kind used for the careful detail work most characteristic of the finer Kuba metalwares. The head and trunk of the anvil were made of copper or copper alloy, though it is uncertain how, that is, whether it was cast or formed by smithing. A pointed iron stem at the other end was smithed, and was encircled by several bands of iron; between these bands, repeating loops of fine copper wire all around the stem produced a bulbous, swelling form rather like the ripe, fruit-laden stalk of a banana tree. This bulbous form would have been the visible base of the anvil when it was inserted into a block of wood.

The variety of materials and techniques in its manufacture already set this anvil apart from those that actually served as smithing tools in finishing work. Several other features tied it more directly to the king and the court. The trunk of the anvil was hexagonal in section, with each of its six faces having the same symbol carefully engraved within it. Not just a pleasing decoration, the symbol, a continuous interlace like an extended figure eight, was the symbol most often associated with Woot. Placing this symbol on the anvil six times gave the object a profound collective meaning and resonance, referring as it did to Woot, the culture hero of the central Kuba groups, reputed to have been the first being at the beginning of the world, who in turn created humans.[63] Seen from any side angle, therefore, Woot graced this precious object and called to mind the stories and ideas that he represented, reinforcing in viewers the temporal depth and cultural unity of the kingdom. At the top of the anvil, engraved on its gently curving, horizontal surface, was the visual and verbal symbol known as *ncyeem*, a potent symbol associated with royalty. While its visual image described the species of beetle called *ncyeem*, this same word held other meanings, which referred to the Creator God and a royal charm.[64] Lacking more specific documentation of how art objects were used on certain occasions, one can only imagine the ways this anvil could have been wielded as a mnemonic reference and accompaniment in public orature. It could well have been the anvil said to have represented the king in local tribute-collecting ceremonies.[65] What is certain is that ceremonial displays of this particular work of art referred to much more than the king himself, and must have struck deep and powerful chords in the viewing audience.

Yet another one-of-a-kind object from the king's treasury was a spectacular sword, reportedly received by the Rev. William Sheppard as a gift of thanks (Fig. 6-4). He was told that it formerly had been used in a yearly ceremony at the capital:

> Once a year thousands from associated tribes [sic] gather to report to Lukenga [the king]. Daily for two weeks they assemble in [a] public square for speeches, singing, and dancing. The Lukenga sits high in his partition surrounded by his family and court. The master of ceremonies lays this long knife before a chief who picks it up, steps out into the circle, salutes the king with the knife and gives his dance as the drums beat and the long ivory horns blow. After this he makes his report.[66]

Apparently the sword had been taken from the treasury by one of the king's sons as he fled the capital after the death of his father. The interregnum period was one of intense competition among candidates and their supporters,

Fig. 6-4: Iron masterpiece from the Kuba royal treasury, Nsheng. Iron blade, spirals and filigree, 45.2 cm long and 3-4 mm thick. William H. Sheppard Collection, Hampton University Museum, Hampton, Virginia. 15.226. Collected by William H. Sheppard before 1911. [Photograph by the author]

and could erupt into armed conflict; hence this son sought temporary protection at the Presbyterian mission station in Ibaanc, and it was in return for this help that the sword was offered to Sheppard.

It, too, was an object unlike any other in the corpus of Kuba artworks. Resembling in silhouette and dimensions the special royal sword held only by the king, the internal structure of the blade itself was unique and visually startling. Bearing none of the basic structural features of a weapon, this sword was made instead to carry visual elements composed in metalworking techniques not seen elsewhere and not usually associated with iron. The length of the blade was divided into four rectangles divided from each other by a horizontal band of braided iron wire. Each of the rectangles was subdivided into a larger and smaller unit, the larger one filled with vertical iron braids and the smaller one filled with circular coils of iron strip. All of these elements were held together firmly, probably by heat welding. The way these circular and interlace motifs were made from lengths of iron wire and strip combined the grace and delicacy of handling fine, pliable thread with the strength and permanence of the iron material itself. Such techniques, known as filigree, are usually associated with wire made of gold or silver. In central Africa, the metals used were copper or iron, from as early as the twelfth and thirteenth centuries,[67] though there may have been earlier examples, possibly also in gold, that are no longer extant. In the case of the Kuba sword, the master smith who made it adapted filigree techniques to an already well-defined aesthetic based mainly on graphic imagery.[68] He manipulated the metal wire not to create independent, three-dimensional forms such as the necklace found in the Upemba burials which was made of plaited strands of copper, but instead to compose elements of repeated pattern, a defining characteristic of Kuba art and design.

Thus there is no question that this unusual sword's worksmanship was Kuba, for the motifs were drawn from the rich visual vocabulary that was shared by most of the kingdom's artisans. Concentric circles and a plethora of braid and interlace motifs can be seen carved, engraved, painted, and sewn onto Kuba products made of all sorts of materials from wood and metal to fiber and beads. Nevertheless, this familiar imagery had probably never been conveyed before in precisely this way. It was one thing for coils and braids to be fashioned out of soft and pliable materials—to produce, for example, the incredible array of beaded belts that served as insignia of office. It was quite another, however, to see strands of iron metal actually coiled or braided together so consistently and evenly, worked in a seemingly effortless way, magically released from the limitations of the material. Hence the visual impact of this sword can be interpreted as something of an astonishment, for it was saying something very old in a witty, totally new manner.[69] That the kingdom's leaders held and danced it regularly in public ceremony could only have added to its renown.

IRONWORKERS AND SOCIAL MOBILITY IN THE KUBA KINGDOM

The various and sundry iron products that have been analyzed and discussed in this chapter reflect in one way or another the competition among men for wealth and favor that helped to create the "Kuba miracle." Subjects of the king enjoyed a higher standard of living relative to their neighbors, and at least some of this economic success can be attributed to the advancements made by ironworkers and other artisans in their work. By the nineteenth century, they were producing manufactured goods that were highly differentiated according to product type and quality. Production of so many commodities and luxuries thus helped to articulate and expand the market for manufactures, and enhanced the flow of trade beyond the kingdom's borders. Artisans also produced for a prestige market within the kingdom, as certain types of objects came to be associated more and more with titled position and social status. Ambition, at least for many free males, was rewarded, and some of those rewards were in the form of elegant costume and equipage. Thus artisanal production also helped to articulate and expand the kingdom's bureaucracy by contributing old and new prestige goods and rarities for show.

Ironworkers themselves showed ambition and competition for favor as they developed new tools, techniques, and products, sometimes importing them from elsewhere, thereby raising the standards of their craft. Ironworkers had made innovations in the designs of smithing tools and various types of standard products, and in the category of masterworks there were smiths who had created some truly admirable works of art. Officials monitored their workshops out of more than pecuniary interest. Yes, there was the percentage of iron goods owed to the government periodically as tribute. But there was also the king's interest in identifying masters and masterworks, so that they might be brought to court to work for him. In this way, ironworkers could hope to receive rewards for furthering their own skills and ambitions, perhaps by earning a title, working among prominent people at the capital, becoming well known in official circles, or having an artwork selected to be preserved in the royal treasury.

Ironworking represented much more than economic activity. In appreciating this, we are a long way toward understanding why the Bushoong oral traditions described iron ore as rather common yet venerable stuff, the spoor of God. And whether or not there ever was a test or battle between leaders on the Iyool plain, the image of clever, witty tricks of smithing now seems very much in keeping with what actually went on in some of the workshops of great Kuba blacksmiths. Oral traditions which featured smithing clichés were not telling literal truths, but were revealing figural ones that were based on shared beliefs and observations. Such clichés were created out of what was known and considered plausible about blacksmiths. The characterizations themselves probably did not pertain directly to specific people and events, but were nonetheless appropriate and acceptable because they corresponded with peoples' experiential reality. Ironworkers could and did epitomize how skilled and enterprising men might achieve success, even royal status, in the Kuba kingdom.

NOTES

[1] Jan Vansina, Field Notes (1953–4), Cahier 32 and Cahier 13, pp. 98–99; and personal communication, September 1997.

[2] See, for example, William Sheppard, "Into the Heart of Africa," reprint of his 1893 address, *Southern Workman* (April, 1895), p. 63.

[3] Vansina, *Woot*, pp. 192–95. See also the maps of nineteenth century overseas trading zones in Jean-Luc Vellut, "The Congo Basin and Angola," in *UNESCO General History of Africa*, Vol. VI, J. F. Ade Ajayi, ed. (Paris: UNESCO, 1989), p. 307; and Vansina, *Paths*, pp. 199, 212–13, and 241.

[4] Vansina, *Woot*, p. 193.

[5] For the high levels achieved in food production, see Vansina, *Woot*, pp. 172–82.

[6] Vansina, *Woot*, pp. 182–83, 187–94; Idem, Field Notes, Cahier 15 (1953–4), pp. 82–83.

[7] Vansina, *Woot*, p. 3. See also Mary Lou Hultgren and Jeanne Zeidler, *A Taste for the Beautiful* (Hampton, Va.: Hampton University Museum, 1993). For a mid-twentieth-century comparison of consumer goods in an ordinary home and the house of an important dignitary, see Jan Vansina, "Trade and Markets among the Kuba" in *Markets in Africa*, Paul Bohannan and George Dalton, eds. (Evanston: Northwestern University Press, 1962), pp. 203–204.

[8] River ores were mentioned in the Ngongo oral tradition of how smelting began, as told by the master smith of the town of Misumba and recorded by Torday. Torday and Joyce, *Notes Ethnographiques* (1910), p. 248. Frobenius briefly described the gathering of riverbed ores and their preparation for smelting near Ibaanc. Klein, *Frobenius*, Vol. II, p. 21.

[9] Vansina, *Woot*, p. 141.

[10] Ibid., pp. 55–56.

[11] Ibid., pp. 55–56, 70, 164.

[12] Torday and Joyce, *Notes Ethnographiques* (1910), pp. 189, 193–94; Torday field notes, "Notes and Drawings" (small format notebook, pagination inconsistent), p. 645.

[13] A brief description of Pyaang smelting suggests a bowl furnace was used. Vansina, Field Notes, Cahier 32 (1954–6), p. 37.

[14] Torday and Joyce, *Notes Ethnographiques* (1910), p. 248. The translation is mine.

[15] Vansina, *Woot*, pp. 270–74.

[16] Torday and Joyce, *Notes Ethnographiques* (1910), p. 235. The translation is mine.

[17] Vansina, *Woot*, pp. 101, 334–35 fn 28, 348 fn 11.

[18] Ibid., pp. 49–52; Vansina, *Geschiedenis van de Kuba* (Tervuren: MRAC, 1963), pp. 111–14.

[19] From the collections in the Ethnology Section, MRAC, Tervuren: Anvils—MRAC 15572, 15573, 15574; Hammers—Acc. # 15575, 15576. Published in Maquet, *Outils de*

Forge p. 36; Pl. IV. For the lexical terms, see Vansina, *Woot*, pp. 271–72. See also the photographs and description in Torday and Joyce, *Notes Ethnographiques* (1910), pp. 189–90, 194. For the wooden base or billot used with the sledge-anvil, see Maquet, *Outils de Forge*, pp. 7–8, 28, 42–43, 46–47.

[20] The emblem portion of the statue is damaged, making it difficult to confirm if the object was indeed an anvil inserted into a cowrie-covered base. See the comparative visual evidence in Joseph Cornet, *Art Royal Kuba* (Milan: Grafica Sipiel, 1982), pp. 94–97.

[21] Torday and Joyce, *Notes Ethnographiques* (1910), p. 195; Vansina, *Woot*, pp. 218, 182–83, 350 fn 47. The reference to Myeel's mother probably refers to his matrilineage, but again, the claim of invention is not to be taken literally.

[22] Vansina, Field Notes, Cahier 32 (1954–6), "Les Valeurs culturelles des Bushong," by Shyáám aNce, insert.

[23] Vansina, *Woot*, pp. 140–41.

[24] Vansina, *Geschiedenis* p. 306; Idem, *Woot*, pp 182–83.

[25] Vansina, *Woot*, p. 175; Torday and Joyce, *Notes Ethnographiques* (1910), p. 195; Klein, *Frobenius*, Vol. II, p. 21.

[26] Jan Vansina, *Le Royaume Kuba* (Tervuren: MRAC, 1964), p. 20; Idem, "Trade and Markets," pp. 204–205.

[27] Vansina, *Royaume*, p. 19.

[28] Vansina, *Woot*, pp. 188, 194–97, 354 fn 91; Klein, *Frobenius*, Vol. II, p. 21.

[29] Vansina, "Trade and Markets," pp. 197–99. However, there was at one time in the past a form of iron arrow blade currency, examples of which were described and collected by Frobenius in 1906. See Klein, *Frobenius*, Vol. II, pp. 25–26, 132.

[30] Kriger, museum notes, technical data from Torday Collection: MOM Acc. # 1908.Ty.155, 1909.5-13.164, .167, .169, .180, .181, .189. The story that these wooden knives were made to be carried at the time of the new moon is accurate insofar as this was a time when certain public events were held; but that does not explain why they were made of wood. See accession documentation for the Torday Collection, and also Joseph Cornet, *Art Royal Kuba*, p. 303.

[31] See Vansina, *Woot*, pp. 3–4. For a striking comparison between Kuba attitudes toward work, wealth, and status and those of their Lele neighbors, see Mary Douglas, "Lele Economy compared with the Bushong," in Bohannan and Dalton, *Markets in Africa*, especially pp. 224–33.

[32] Rev. Motte Martin, "The Bakuba people, as seen in a trip to Lukengo's Home," *The Missionary* (July 1904), p. 340.

[33] Vansina, Field Notes (1953–4), Cahier 13, pp. 98–99; Idem, *Royaume*, p. 108.

[34] It is structured like a hoe blade, but documentation information states that the same kind of handle was used for adzes. Sheppard Collection, HUM Acc. # 11.218. Torday reported that a type of hoe was an emblem of office for titled royal women who were in charge of the women at court, but his drawing of it does not show any detail for the form or surface of the blade. Torday and Joyce, *Notes Ethnographiques* (1910), pp. 58, 72. The adze reportedly used for ceremonial purposes was only known as *ikééng* in Bushoong, related to Tetela and Nkucu terms for axe, and the Luba Kasai term for adze. Vansina, *Woot*, p. 273. However, Luba Kasai shows the meaning hatchet, as do the Mongo Lokalo and Losikongo. In Nyanga (J43), the term *kakéngé* means hoe. The combined data and distributions suggest that this is a recent borrowing through trade of the object, a cutting blade that could be mounted in different ways or ornamented for different uses.

[35] Torday noted that at the time of his visit there were certain types of axes that were among the insignia of the *cikl*, the *ipaancl*, and the *nyimishoong*. See Torday and Joyce, *Notes Ethnographiques* (1910), p. 53 and figs. 36 and 37. Vansina considers the first two of these titles to be very old, and the words probably derived from loans of Luba, and Luba Kasai or Songye, origin. *Woot*, pp. 120, 124, 310–11. Just as the titles underwent change over time, the insignia did too, and it should not be assumed that axes were well established as insignia. It is more likely that they were recent. Sheppard collected a copper axe, and stated that it was given to him by one of the king's sons. It was probably from the treasury. He described it as a sign of authority carried by messengers to identify them as representatives of the king. HUM Acc. # 11.190. The ceremonial adze collected by Sheppard (# 11.220) was not described as any kind of insignia or emblem, though there have been some references to adzes as such for the holder of the title *shesh*, and the holder of the Pyaang title *ipaancl*. See Vansina, *Woot*, pp. 120, 273, 310; and Idem, Field Notes (1954–6), Cahier 32, p. 43.

[36] See the illustration of "Bakuba Weapons" in the 1880s, #3, "ikullo." Von Wissmann, et al. *Im innern Afrikas*, p. 252.

[37] Kriger, museum notes, technical data from eight specimens collected by William Sheppard, the first foreigner to visit the Kuba capital (in 1892): four now in the Sheppard Collection, HUM Acc. # 11.191, .192, .200, and .208; four others were originally collected by Sheppard, two of them sold to the Redpath Museum in Montréal, Acc. # 3773.01, and two given to the Rev. T. Hope Morgan and now in the ROM, Acc. # HA 2454 and HA 2456. I examined another ten specimens in the Torday Collection, MOM Acc. # 1909.5-13.163 - .169, 1908.Ty.155, and 1907.5-28.349 and .350; and out of 178 surveyed in the MRAC, I examined fourteen selected specimens, # 16030, 40106, 16024, 16027, 21994, 21995, 16015, 16016, 16019, 16020, 16021, 16022, 16023, and 17507.

[38] Accession documentation, Sheppard Collection, HUM.

[39] Accession documentation, Sheppard Collection, HUM.

[40] These comments are based on my own assessment, made after viewing several thousand knives, swords, and hatchets from the Zaire basin, and also on the comments of two blacksmiths I interviewed while viewing the Hope Morgan (Goad) Collection of metalwork from the Congo Free State, now in the ROM in Toronto. Transcript of interview, David Norrie and Daniel Kerem, blacksmiths, Toronto, 10 April 1989, pp. 5–6.

[41] *Ikul* knives were made by blacksmiths in Ngeende, Bushoong, Pyaang, Coofa, and Kete communities. Vansina, *Royaume*, p. 20. The lexical term can be traced back to southern Mongo. Idem, *Woot*, p. 272. Its distribution and the distribution of manufacture together suggest that it is the oldest of the prestige knives known. A less complex version of the stepped midrib can be seen on a type of knife made in Songo Meno smithies, further supporting this suggestion. See #5 in Von Wissmann, et al., p. 252 and Torday Collection, MOM Acc. # 1909.5-13.171.

[42] See "Ilohno," #3, in illustration "Bakuba Weapons." Von Wissmann, et al., p. 252. This description and discussion are based on my museum notes and technical data from a survey of 56 examples in the Ethnology Collections of the MRAC, and my analyses of the following other examples: Torday Collection, MOM Acc. # 1909.5-13.177-.184, 1909.Ty.664 and .665, 1909.Ty.490-.494. See also the examples noted by Frobenius in Klein, *Frobenius*, Vol. II, pp. 82 #104, 84 #111, 133 #208, 139 #218 and #219.

[43] Surface patterning features are arbitrary, and are therefore most likely to be cultural. Kuba patterning was conventionally done on the right half of each blade face; on the *iloon* swords, the patterning was done on the left half of each blade face. The latter convention was a major defining feature of what I call in my technical analyses "Group

C," distributed in a geographical area to the east of Kuba. Kriger, "Museum Collections," pp. 137, 139–40.

[44] See Vansina, *Woot*, p. 274. *Kilonda* means "iron" in Luba-Shaba and in Hemba; it means parade hatchet in Songye, while the Songye term *bilonda* means "iron."

[45] Vansina, *Royaume*, p. 116.

[46] They are structurally unlike the famous war swords of the Kete and Salampasu, south of the kingdom, and are unlike any other knife or sword known to have been made in the nineteenth century. It appears that the blacksmiths' workshops of the Kuba kingdom successfully appropriated the skills and techniques for making them.

[47] Vansina, personal communication, 18 February 1993; Idem, *Royaume*, p. 147.

[48] Kriger, museum notes, technical data from Sheppard Collection, HUM Acc. # 11.194, .195, .197, .202; and Torday Collection, MOM Acc. # 1909.Ty.666, .668, 1908.Ty.239, 1909.5-13.187, .188, .190.

[49] Kriger, museum notes, technical data from MOM Acc. # 1909.5-13.189 and .191.

[50] Kriger, museum notes, technical data from Sheppard Collection, HUM Acc. # 11.207 and 23.204; Verner Collection, FM Acc. # 88180; and Torday Collection, MOM Acc. # 1909.5-13.170 -.176. I surveyed 17 examples of the wooden knives in the Ethnology Collections of the MRAC. See also examples noted by Frobenius in Klein, *Frobenius*, Vol. I, p. 203 # 525, p. 209 # 546; Vol. IV, p. 224 # 408; and the illustration in Joseph Mertens, "Les BaDzing de la Kamtsha," *Mémoires. Institut Royal Colonial Belge, Section des Sciences Morales et Politiques* 4 (1935), p. 191.

[51] This may explain the presence of an otherwise unremarkable example of *ikul imbaang* that was given to William Sheppard as a valuable gift in 1892. The blade was made of iron, and was not particularly impressive technically, unless it had been achieved by copying from an imported model. Sheppard Collection, HUM Acc. # 11.207.

[52] Kriger, museum notes, technical data from Sheppard Collection, HUM Acc. # 23.204.

[53] Particularly the use of an open sand mold, rather than the clay molds used south and southeast of the kingdom. See Torday and Joyce, *Notes Ethnographiques* (1910), p. 194; and Maquet, *Outils de Forge*, p. 94.

[54] Kriger, museum notes, technical data from Torday Collection, MOM Acc. # 1909.5-13.192.

[55] Klein, *Frobenius*, Vol. II, p. 138 # 216 and 217.

[56] In some cases, there were probably blacksmiths who achieved renown for their skills and were then summoned from provincial towns or villages to work in the workshops in the capital. That such cases could, and probably did, occur is suggested strongly by the admittedly apocryphal story told by Bushoong blacksmiths about the first Bushoong smith. He reputedly rose from being a commoner to a smith who worked at court; henceforth his entire family became a royal family because of his blacksmithing prowess. Vansina, Field Notes (1953–4), Cahier 13, pp. 98–99.

[57] Less frequently mentioned was another such famous smith, referred to as Pyeekol, who was also believed to have excelled in metalworking. Vansina, *Woot*, p. 218.

[58] One example is an ornamental axe, its blade made entirely of copper, that was given to William Sheppard. It was said to have been carried by the king's messengers while on official errands. Sheppard Collection, HUM Acc. # 11.190. It is unusual also in its structure, for it has a midrib along the length of the blade, a feature not typical at all of hatchet blades. Oral testimonies claimed that there had been a metal anvil decorated in cowries that represented the presence of the king, and that was sent with his officials

to villages in the kingdom where it was displayed in public tribute-collecting ceremonies. Cornet, *Art Royal Kuba*, pp. 94–96.

[59] Torday and Joyce, *Notes Ethnographiques* (1910), pp. 25, 37; Vansina, *Woot*, pp. 67, 182.

[60] See the photograph and cursory description in Cornet, *Art Royal Kuba*, pp. 288–89.

[61] Adriaan Claerhout, "Two Kuba Wrought-Iron Statuettes," *African Arts* 9 (1976), pp. 60–64, 92.

[62] Formerly part of the kingdom's treasury, then in the collections of the IMNZ, Acc. # 70.6.280, 28.9 cm high. See the photographs in *Sura Dji. Visages et Racines du Zaire*, (Paris: Musée des Arts Décoratifs, 1982), pp. 82–83; and Cornet, *Art Royal Kuba*, pp. 289–91. Its present whereabouts are unknown; the IMNZ collections were sacked during the collapse of the Mobutu regime in early 1997.

[63] For the tales of genesis, see Vansina, *Woot*, pp. 30–40.

[64] Ibid., pp. 317, 379; Cornet, *Art Royal Kuba*, p. 179.

[65] See fn 58.

[66] Accession documentation, Sheppard Collection, HUM.

[67] See Kriger, *Ironworking*, pp. 137, 370.

[68] For an example of how well developed this aesthetic was, see Dorothy Washburn, "Style, Classification, and Ethnicity: Design Categories on Bakuba Raffia Cloth," *Transactions of the American Philosophical Society* 80, Pt. 3 (1990).

[69] A related knife, made in the shape of an *ikul* knife, and filled with coils of iron strips, was probably made by the same hand. It was reputedly worn by the king in the nineteenth century, and was also a gift to Sheppard from a member of the royal family. Kriger, museum notes, technical data from Sheppard Collection, HUM Acc. # 11.198.

7

Markets for Prestige along the Middle Zaire

Botúli oa efolóka, otúlelaka balé nd'otónga nkulá.
The blacksmith, an only son, hammers out the knives for making large families.
Is'ék'atúli ooo !!!
The father [best] of blacksmiths !!!

—Mongo[1]

This second case study examines some of the social and economic values of iron in a single town, a town founded by blacksmiths. The chapter begins with a brief history of the town's settlement and a discussion of its heritage as a mining, smelting, and smithing center. I then identify several prominent blacksmiths who generated and accumulated iron wealth, showing that this form of wealth in particular gave them clear advantages over other men in buying slaves and in acquiring wives through bridewealth payments, transactions that were conducted with iron currencies. Hence as "big men" they came to be known as founders of "lineages," and sometimes of entire communities. A major portion of the chapter traces the histories of five types of prestige knives, which together demonstrate here as in the previous chapter the stratification of society, in particular the divisions among the male elite, and how men publicly displayed their social positions. The chapter ends with examples of how the social mobility of certain nineteenth-century ironworkers did not end at their deaths, but continued through time as they became remembered historical figures.

I reiterate here (with some slight variations) the main argument in Chapter 6, which concerns the relation of ironworkers to political authority. In this case, there was no centralized governmental administration and hence no formal system of specifically royal patronage. Nevertheless, I argue that centralized government was not a necessary precondition for product innovation, elite patronage, technology transfer, and the social advancement of ironworkers to take place. By focusing on a single town, I bring to light more detailed evidence about individual masters of the past century. As was true for the Kuba kingdom, it was blacksmiths especially who were known to have been wealthy and vital to the precolonial economy, and their most complex and elaborate luxury products were valuable markers of social and political position. In Lopanzo, metal products permeated all sorts of social institutions—facilitating marriages, sanctioning judicial procedures, articulating social relationships, and commemorating local history. And here, too, individual and collective treasuries offered proof that ambitious master smiths of the past had managed to achieve greatness and legendary status.

IRONWORKING AND THE HISTORY OF LOPANZO

The town of Lopanzo was an important iron center in the nineteenth and early twentieth centuries, during which time it served as a source area for ore, for smelted iron, and for iron products. Now, as in the past, the town is populated by settled agriculturalists and their Twa clients, each with their own subgroups of houses which they refer to as lineages, and all speaking the Mongo language of Lokonda. Based on what is known from genealogies and other historical data collected earlier in this century, the founders of Lopanzo were part of a larger movement of Ekonda populations. Their departure southward from their earlier homes along the Ruki River to their present location north of Lake Mayi Ndombe is estimated to have begun no later than between 1660 and 1695.[2] This estimate corresponds with the genealogy recounted by the spokesman for the senior house of Lopanzo, which presently includes six ascending generations before c. 1940. Following Sulzmann's proposed forty-year average distance between generations, that would time the reputed departure of founder Mbongo Ilomoto from the town of Longa on the Ruki to no later than 1700.[3] After an undetermined period of settlement in the Butela area, he and his followers established themselves near the present site of Lopanzo (see Map 2).

The emigrations of Ekonda houses and communities from the Ruki toward the south is explained in their oral traditions as having been forced on them by Nkundo groups sweeping in from the east and northeast. This was not an invasion of foreign strangers, for Lokonda and Lonkundo are close to each other in the Mongo language family.[4] Nkundo dominance in their relationship was expressed genealogically: the Nkundo ancestor, Bongo, was the elder of Mputela, ancestor of the Ekonda. Smelting, smithing, weaving, and certain cultural practices were claimed to have been acquired by the Ekonda from the Nkundo, though this probably refers only to certain transfers or changes, not complete introduction. For example, lexical terms for smithing tools and the tools themselves are the same in Lonkundo and Lokonda, the anvil *njondo* and the hammer *bosákwá*; both Nkundo and Ekonda ethnic groups distinguish between smelting and smithing; and there have been specialists in the past in the two ethnic groups who did both.[5] But there remains a question as to just what techniques Ekonda ironworkers actually learned from Nkundo masters.

Indirect evidence suggests that some of them learned to smelt. Lopanzo historians claimed that the knowledge and techniques for smelting iron ore were transmitted to their forebears by Nkundo ironworkers. Before that time, Ekonda blacksmiths apparently had periodically invited Nkundo smelter-smiths to come and work for them for a fee.[6] This was not likely to have been a simple hiring process, but probably involved the active assistance of Ekonda workers on the smelting team, who by paying their share of labor and supply costs were able to earn a share in the bloom. One may infer from this account that the transaction was arranged by Ekonda blacksmiths, for otherwise these communities would simply have imported iron products rather than bringing in smelters, and they would have to have been able to offer labor, including semiskilled assistants. The Waya section of the Ekonda, who themselves only did a limited amount of smithing, reported what appears to have been a common pattern of blacksmiths supporting, and sometimes initiating, visits to their communities by master smelters. Waya elders recounted a time in the late nineteenth century when smelters from the Lioko section came to their villages to set up smelting and refining operations.[7] In this case, however, the technological knowledge of smelting was not transmitted completely to the host group.

It is not clear whether or not Ekonda ironworkers had learned smelting before their emigrations began. But since Lopanzo, at least, was founded near an iron ore deposit, it is probable that they had. There may have been a time of peaceful coexistence between Nkundo and Ekonda communities before then. There may have been alliances or client relations between their leaders which continued even during the emigrations. These alliances could have facilitated the contacts and arrangements for the transfer of smelting technology to Ekonda smiths. On the other hand, Ekonda smiths could have learned to smelt from other ironworkers in the Ruki area. Numerous slag heaps and furnace remains at the Zaire-Ruki confluence were attributed to Sakanyi smelters, who inhabited the area of Mbandaka before they, like many Ekonda, were displaced.[8] Some basic features of this furnace design appear also in the furnaces constructed in Lopanzo for smelts commissioned in 1974 and 1989. A systematic comparison of furnaces and other material remains at both sites will shed more light on the history of iron smelting in the Ruki River area before the nineteenth century.

The historical picture south of the Ruki is not a simple one either, for it appears that the Ekonda were not necessarily the first miners and smelters of iron in the lands north of Lake Mayi Ndombe and east of Lake Tumba. There were three sections of the Ekonda ethnic group who produced significant amounts of iron, leaving behind heaps of slag from smelting furnaces and refining forges. Elders among these sections—the Besongo, Lioko, and Liombo subgroups—claimed that they had smelted iron up until about 1915, and identified slag heaps with the work of their forebears. Occasionally, however, they attributed one or another heap to the former inhabitants of their territory whom they themselves had displaced. The same was true for abandoned mines, some of which apparently had been worked by others before the arrival of the Ekonda immigrants.[9]

Lopanzo was founded in the area identified as the Besongo section, yet it is the only ironworking center of any of the Ekonda subgroups in which any research on iron production has been carried out. It was founded by three ironworkers, each of whom is remembered as the ancestor of a major house, the "eldest" house having rights over the land and nearby iron ore deposit. The present layout of the town dates from the early 1950s; at a former town site nearby, slag heaps confirm that smelters indeed had worked with some regularity at least around the late nineteenth and early twentieth century. The last smelt took place in the 1930s, and so it was a process witnessed by and within the living memory of elders who performed a demonstration smelt for researchers in 1974.[10] Smithing continues to this day, although in 1989, when another smelt was commissioned, there was only one active smithy in the town.[11]

SMELTING

The reconstructions of iron smelting processes in Lopanzo in 1974 and again in 1989 illustrate the difficulties surrounding what has come to be called "ethnoarchaeology." The paramount question that arises is just how much of what is being done today as a "reconstruction" represents what was done in the past and what is strictly a phenomenon of the present. In Lopanzo, this question was particularly relevant since the reconstructions were based on the memories of men who had not been practicing smelters all their lives but were, as youths, assistants to or observers of master smelters. The experiential base had been ruptured, and it was not clear how much of the technology was recovered and remembered, or how much reinvented on the spot. In 1974, two master smelters supervised the smelt: the master smelter from the Nzelika house, Nyekuli Bana; and a master diviner-smelter from the town of Lokakau, near Penzwa, just east of Lopanzo.[12] Nyekuli Bana (c. 1900–1980) had never carried out a smelt on his own before 1974, but in his younger days he had assisted his father, the renowned master smelter Nsabonzo. Perhaps because of his lack of experience, perhaps because he believed the cooperation of a diviner-smelter would insure success, Bana requested the assistance of another master in 1974. By 1989, both of the specialists who directed the 1974 smelt had died. Undeterred, the Nzelika house designated a senior elder to hold the position of master smelter, though it was an honorary one since he had little direct experience of the working process. Added to the team were the two sons of the former diviner-smelter, who were summoned from Lokakau to be in charge of the ritual aspects of the smelt.

A simple but erroneous solution would be to dismiss these kinds of reconstructions entirely. The *capita* or mayor of Lopanzo, for example, turned the 1989 reconstruction into an occasion for the playing out of house and town politics through the selection of managers and workers and through the negotiation and distribution of payments for the work. Labor requirements were expanded to meet current economic conditions and social needs, and the lack of a clear technical master exacerbated the fractiousness of social relations in the process. But the past was not entirely absent. The smelt was a challenge and an opportunity for elder men in Lopanzo to demonstrate publicly the pride and privilege associated with the skill of smelting iron, and to pass on to the younger generation what little remained of the body of technical knowledge built up over the years by their forebears. As such, the smelt was very much a present day affair in which there were some few elements remaining from the past.

Based on local conditions at different times, master smelters, as managers, constructed and reconstructed labor processes around the four major technological requirements of charcoal, ore, furnace, and the smelt itself. Charcoal was made primarily from two species of tree; the preferred one, *bosénga*, was common in secondary forests and recognized as a wood that produced a dense, superior charcoal, capable of generating high heat and used both for smelting and for smithing.[13] Charcoal manufacture in the past was described as having required only the smelter and his Twa client, the client doing all the manual labor. For the 1989 smelt, however, four teams of four Twa men each were organized, presumably to insure that the cash compensation was more widely distributed.[14] The teams built tall holding structures for the wood, burning it slowly for a day to remove the sap. Later, they reputedly covered the entire mass with leaves and clay for the slow roast which, after another day, resulted in charred wood, the charcoal fuel for the smelt. The process was carried out in the forest, next to the felled tree. That blacksmiths had continued to work in Lopanzo probably accounts in large part for the continuation of knowledge about the charring process, and the choice trees for making it. In this instance, the technological knowledge had been maintained over time, but the labor process was adapted to current conditions.

According to the remembered past, mining was done by individuals or small groups in need of iron for making specific implements or for bridewealth. Permission to extract ore from certain deposits had to be sought from the house with rights over the mine, which in the case of Lopanzo was the "eldest," the Nzelika house. The Lopanzo mine, just a short distance away from the town, showed evidence of having been exploited regularly over time, though we were unable to make estimates of the approximate total yield. Ore was available at relatively shallow depths of about two meters. Tools used for the extraction of ore were hoes and wooden picks; machetes were used to break up the blocks into smaller pieces for sorting. Elders seemed particularly keen on teaching the younger men and boys how to recognize the better ore from richer veins by its color.[15] Again, however, there were many more workers in 1989 than ever would have been the case in the past. About forty workers took part in various tasks associated with the mining and sorting of ore, preparing about 400 kg of it, or roughly four times the amount of ore used in the actual smelt.[16]

The laterite ore, being hydrated, had to be roasted in preparation for the smelt. This roasting process would have the effect of reducing the burden of the furnace by ridding the ore of water and carbon dioxide, for example, and by increasing its porosity.[17] However, it is not clear what had been understood locally as the benefit gained from the investment in this additional step in the process.[18] After the roasting, the ore was once again broken up into smaller pieces and sorted. Prepared ore averaged slightly smaller dimensions than the charcoal, the latter measuring about three by three by two cm.

On the day of the smelt, the master smelters and their assistants constructed the furnace in the same location as where the 1974 smelt took place. It was an oval bowl furnace, about 110 cm long, seventy-three cm wide, and twenty-five cm deep, and had a slag-tapping tunnel leading from one side of the bowl to a large observation pit. One of the actively employed blacksmiths constructed a clay tuyère on site, using prepared clay molded onto a woven fiber frame; the wooden bellows which had belonged to the former master smelter, Bana, were used, one with two and the other with four drums. Another bellows was borrowed from the master blacksmith Lotoy. My attempt to keep track of temperatures in the furnace was only partially successful. The thermocouples placed in the bowl could not be fixed at a permanent point for constant measuring in the reducing area, and there was much shifting around and mounding of ore and charcoal, using wooden paddles, so that those temperatures that were recorded are for unspecified locations inside the furnace. I kept a log of the smelt, including approximate weights of the charges and their timings; slag runoff occurred during the 1974 smelt, but not in 1989.[19] The entire smelting process lasted about five hours and yielded only a small bloom with a poor yield of iron, indicating that the lack of ongoing practice had resulted in a serious loss of technical knowledge, even since 1974.[20] All in all, the team of smelters came to about thirty men, including the four blacksmiths active in Lopanzo at the time; in the past, however, the master smelter normally worked only with one or two assistants in addition to the bellows operators (two or three men per shift).[21]

Several questions arose about the technology, based on some of the procedures we witnessed and the equipment used during the smelt. Judging from my log, it appears that the charcoal to ore ratio of approximately six to ten in the Lopanzo smelt was unusual, and perhaps an indication that the smelters did not know what they were doing. However, this ratio would have been sufficient for producing a low carbon bloom, and laboratory experiments have shown that ratios of one to four are possible.[22] Moreover, it is unclear what the ratio was in the reducing area of the furnace, since the charges were dispersed throughout the bowl and were periodically mixed and shifted around with wooden paddles. Also, the large size of the clay tuyère at first appears to have been inefficient, evidently diminishing the power of the forced draft. However, the large tuyère allowed the master smelter, bellows operators, and others to view and monitor the interior conditions of the furnace, in a way similar to what is done in modern foundries. This was undoubtedly the means by which it was discovered part way through the smelt that the tip end of the tuyère was blocked; without a wide orifice at the other end, such an occurrence might have gone unnoticed. One drawback was that this single tuyère meant that the reducing area of the Lopanzo bowl furnace had limited dimensions, and was therefore not capable of producing large blooms.[23]

As laboratory experiments confirm, smelting iron below its melting point in small scale furnaces is a high-risk procedure and experts are not assured of repeated success.[24] An appreciation of the technical difficulties, even for a master, was built into the operations witnessed in Lopanzo. The entire sequence, from the selection of the site to the smelt proper, was surrounded by ritual procedures and invocations to forebears which were designed to insure a successful outcome. There were simply too many of them to describe here, but similar kinds of rituals have been noted elsewhere in connection with ironworking and other important and risky technical processes.[25] In Lopanzo, at least, they contained elements of local history, not only literally in the calling out of names of master smelters of the past for their help and inspiration, but also in the historical content of consecration ceremonies for the smelting site. The current land chief, that is, the eldest male of the senior branch of the senior house in Lopanzo, joined with his Twa counterpart in performing a benediction to bless the site and insure success.

The importance of proper ritual preparation and guidance was evidenced by the requests, in 1974 and 1989, for assistance from a diviner-smelter. Although diviners, like smelters and smiths, were and are specialists, certain individuals combined professions and were diviner-smelters or smelter-smiths. Such instances indicate not the absence of specialization but the ways men chose to diversify their expertise over the life span of their careers. The diviner in 1989, Ndongo Bokoti, was, he claimed, like his father, summoned to Lopanzo to manage the ritual side of the operation, since the master smelters in Lopanzo were themselves not diviners.[26] However, there are indications that it was also his smelting expertise that was desired. The designated master smelter in Lopanzo in fact had no smelting experience, and his lack of competence came to be seen as a threat to the entire operation. Several days before the smelt, Ndongo was asked to assume the role of master smelter as well, much to the consternation of many in the town. Although this particular incident was prompted because the smelt was a reconstruction, it suggests that smelting teams in the past were probably composed in a variety of ways, depending on recent experience and masters' confidence in their chances of success.

Numerous prohibitions were in force during the preparations for and carrying out of the smelt. Women were most consistently targeted as forbidden, although young children, too, were kept from coming within view of certain stages in the process, especially when the furnace was working. There was disagreement about whether women were allowed near the site of charcoal manufacture, but no one disputed the existence of firm restrictions against women approaching the smelting furnace site.[27] Twa men were allowed at the site, but were not included directly in the work team during the actual smelt. In my view, these prohibitions undoubtedly served several aims, but should be distinguished from other behavioral prohibitions, such as those exercised by men to prevent failure of a smelt, e.g. against sexual relations with women or against contact with menstruating women. One of the effects of restrictions keeping women away from certain work sites was the maintenance of control over the knowledge and expertise by master blacksmiths and their chosen assistants. But there was another important reason why men would be intent on preventing women from acquiring any knowledge of ironworking. Since iron was used before the colonial era as a currency and in bridewealth payments, for men to monopolize iron meant that men, especially ironworkers, could monopolize marriage arrangements and the circulation of women.

BLOOMS, BARS, AND BRIDEWEALTH

Iron, in the form of semifinished and finished goods, circulated in currency and bridewealth networks along the middle Zaire and its hinterlands in the nineteenth and early twentieth centuries. Bridewealth remains a poorly understood topic in African economic history.[28] It is sometimes described in terms such as "gift giving" and the formation of family or lineage alliances, other times it is likened to the commercial purchase of women for their productive and/or reproductive labor. Issues concerning just what bridewealth transactions were, and how distinct or overlapping were the so-called economic spheres of gift and commodity exchange will not be discussed at length here, mainly because precise historical data for the nineteenth century are still so sparse. For example, it is still poorly known just what forms of iron were used in bridewealth transactions in the middle Zaire area over time, and the changing forms of iron that circulated as currency or served as currencies of account.

Bridewealth transactions were a complex series of payments from the family and house of the husband to the family and house of the wife before the marriage and throughout its duration. Colonial reports from the Lake Leopold II District (Lake Mayi Ndombe) in the 1920s and 1930s provide evidence of what elders in sample communities remembered as proper bridewealth transactions in the late nineteenth century. Then, marriages were sometimes arranged by choice, but most frequently by agreements between families, sealed by the transfer of preliminary gifts to the future bride and members of her family. This early stage in the process was called *likulá*, or arrowhead, a term that refers to the historical use of arrowheads in Mongo-speaking areas as titles of proof for various transfers of responsibility, including marriage. The wife went to live with her husband permanently only after the main bridewealth payment (the third of four) that had been agreed upon was finally and publicly made.[29] A dowry, provided by the wife's family, consisted of household utensils, domestic animals, and other items essential for the household. But this was not the end of it for the family and house of the husband. Payments and gifts to the wife's family continued with visits, in times of sickness, for example, and at the birth of each child. Children whose ties to the father's house were not firmly established by the requisite bridewealth payment could be claimed by the wife's family. The ideal marriage was characterized as one in which the bridewealth was never finished, and the process has been likened to a perpetual debt.[30]

Bridewealth payments were crucial business for the family and houses of both parties, not only in circulating and redistributing wealth but also by presenting an opportunity for individuals to assume certain positions and responsibilities within each house. Above all, they were important transactions that served to concentrate wealth in the hands of men, while at the same time providing men with occasions when they solidified or modified their alliances with one another. The husband's father, leader of his house, and/or maternal uncle were potential contributors in making payments; recipients were primarily the father, head of her house, and maternal uncle of the wife. Hence such transactions involved at least four different residential communities, namely those of the husband's father, the husband's mother's brother, the wife's father, and the wife's mother's brother. What was received as bridewealth was often then paid out to another house as the marriage payment on behalf of a boy in the wife's family.

Bridewealth was not only a continuous process, but it involved an assortment of different types of goods and currencies, mostly metals. Payments at various stages could include bar iron, copper alloys, and blade currencies, along with arrows, lances, knives, slaves, guns, goats, sheep, and dogs.[31] The largest portion of the payment was usually in the form of currency, though the assortment included a variety of fungible items ranging from those associated with local exchange of goods and foodstuffs to those tied to regional and external trade networks. Clearly, too, the forms and amounts of payments varied from place to place and over time. In Lopanzo, for example, blooms of iron had served in bridewealth payments in precolonial times; after around 1895, when brass wire currency of the early colonial regime reached Lopanzo, so did copper.[32] Long bars of iron imported from overseas also served in bridewealth payments, and may have been at least partly responsible for a decline in the value of locally produced blooms that was remembered from around the 1930s.[33] The relatively low cost of imported bar iron from overseas was cited as a cause of the inflation of bridewealth amounts in the upper Tshuapa area, which hastened a decline in local smelting.[34]

Ironworkers, along with other wealthy men, were therefore in an advantageous position for making bridewealth payments, for they were able to produce or acquire more readily and cheaply the types of goods and currencies used in those payments. Master smelters in Lopanzo, for example, accumulated iron by regularly retaining a share of the blooms they produced on contract for others. One of the famous Lopanzo smelter-smiths, Nsabonzo Engala, whose peak productive period was ca. 1880–1910, was remembered as a wealthy man, a "founder of a lineage" who had had twelve wives, nine of whom bore him children.[35] He was recognized as a man who had been powerful enough in his house to raise political tensions, which were acknowledged indirectly in oral testimonies. Nsabonzo might have moved his dependents to found a new community, but he did not, and so was described as a man who deliberately chose to respect the seniority of the elder branch of his house. He had good reason to accept the *status quo*, for his wealth as a master smelter-smith was augmented by his favorable position as a member of the house with rights over the local iron mine. Other ironworkers, blacksmiths who did not have smelting experience, could accumulate iron in trade or in fees. Moreover, a significant portion of bridewealth payments was in the form of their own products of the smithy, finished iron goods. Hence both smelters and smiths were among those wealthy notables who cornered the market in local women by beginning bridewealth payments for very young girls, even infants, paying relatively larger sums than other men.[36] In this way, then, successful ironworkers could increase the scale of polygyny, altering dramatically the structures and sizes of their communities. No wonder that elders in Lopanzo in 1989 still associated their local deposit of iron ore with the origin of their mothers, that is, with the bridewealth payments that sealed their mothers' marriage contracts and signaled the time for them to take up residence in their husbands' town.

THE SMITHY

Townspeople described apprenticeship in blacksmithing as normatively the prerogative of boys with familial ties to smiths, and ongoing training was undoubtedly necessary for the ones who hoped to attain mastery of the craft. Men and women alike insisted that women were never involved at all in any of the ironworking processes. Twa men, also prohibited from training in the occupation, were able at least in some cases to learn smithing but not smelting.[37] It was acknowledged without hesitation that it was not unusual for boys outside smithing families to become smiths. They did so by exhibiting talent—having been "born into it," having a "gift from God"—and becoming skilled independently through persistent observation and practice.[38] Master smiths tested the basic competence of their apprentices by having them manufacture an arrowhead, and if the level of skill was deemed sufficient, the new smith paid a fee and was then allowed to set up his own smithy.[39] But independence was not the equivalent of mastery. Mastery was attained not by one's "lineage" or house position, that is, not by seniority or relation to the master blacksmith, but rather was based on a judgement by experts that a smith had acquired the highest level of skill. Mastery was visible to the attentive eye, one could see who had reached it by the superior quality of his work.[40]

As was common throughout the region, the most important tools were the anvil and hammer, which identified their owner as a smith. In Lokonda, they were called *njondo* and *bosákwá*, respectively. Some were still being guarded as treasures by house leaders in Lopanzo, long after their owners' deaths.[41] The smithy's master in 1989, Lotoy Lobanga, claimed that in 1961 his own master, Ndzou, using imported iron, made the two anvils and at least four hammers that were in use at the time. He also claimed that the tools were for anyone to use, but further questioning revealed that women were prohibited from using or even touching them, and that Twa men could use them but only under supervision on the premises.[42] In practice, the three apprentices working with Lotoy were all related to him, and each kept his own hammer and brought it to the smithy every day.

Blacksmiths could make most iron goods with just the basic tools of anvil, hammer, and bellows. Plain blades, such as hoes, hatchets, adzes, and razors, required no additional equipment. Some of the harpoons, arrowheads, and lance blades, however, were split with a chisel to form barbs, a technique that improved the efficacy of the blade. Splitting, cutting, incising, and punching techniques were used for manufacturing luxury products, the elaborate lances and knives carried for public display. The complete toolkit of a master smith, therefore, consisted of anvil, hammer, punch (*bobókwá*), and chisel (*losélu* or *loséno*), the latter also doubling as a burin. Files were used for finishing, Lotoy's being an imported one (*bokósa*). In past times, however, a locally made file had been used called *losio*.[43] Another tool that was considered essential was an edged instrument, *bokóti*, for making scores on the surfaces of the anvil and hammer to prevent the worked iron from slipping.[44] These scores imparted to the surface of the product a slightly hatched texture, which can be seen on many museum objects that were collected all over the central basin in the nineteenth century.

The hammer, similar to the one used in the Kuba kingdom, was extremely versatile and practical in skilled hands. A blacksmith could turn out a socketed arrow blade in about ten minutes. He worked with the side of the hammer to draw out the rod of iron, then used the hammer's end to hit the surface repeatedly in order to flatten it. After reheating the iron in the fire, he worked again with the side of the hammer for drawing out and upsetting the iron mass. Soon he had formed a socket, inserted a small rod into it for use as a handle, and then he cut off the excess iron at the other end. He formed the blade out of the remaining mass of iron, drawing it out and flattening it gradually into a point. To make an ogee midrib, he placed the blade along the edge of the anvil, hammered it, and then turned it over and repeated these strokes on the opposite side. Blade edges and sockets were finished by filing and polishing. Smiths often worked on at least two or three products at a time, each one at a different stage of manufacture.

I observed three other techniques at Lotoy's smithy that deserve special mention. On only one occasion was any welding done, and it was in the making of an eyed needle. Although the iron source was imported, hence lacking slag inclusions for easy hammer welds, the smith did not use flux in this or apparently in any other case. Lotoy insisted that nothing other than heat was ever used in welding.[45] Quenching was a technique reserved only for the making of harpoons.[46] Lotoy explained that they used the quenching technique to make the iron stronger, and that it was called *bóló* in Lokonda, "very strong."[47] However, I purchased one of the quenched harpoons, had it analyzed, and its microstructure indicates that temperatures high enough for quenching were not reached in the forge, and that the carbon content of the iron was not high enough for the quenching to have made a significant difference in the hardness of the iron.[48] It is not clear whether this was the result of insufficient training on the part of the smith working under Lotoy, or if the skill in judging temperature for quenching had been lost or was not clearly known in the first place. The third noteworthy technique I observed was when Lotoy blackened a ceremonial knife blade by rubbing it while still hot with the freshly cut cross section of a sugar cane stalk. When asked why this was done, Lotoy stated that it was done to make the blade a darker black, so that it would look beautiful.[49]

The four smiths at work at Lotoy's smithy had attained various levels of expertise, limiting the output of each to a certain range of products. Arrowheads were the most frequently made, worked up to sell to whomever needed them, the market being relatively constant for them, especially among Twa hunters.[50] All of the four smiths at work manufactured arrowheads, lance blades, harpoons, tools for plaiting hair, and other utilitarian items. Prices for products of the smithy, and the daily output I observed, indicate that even recently blacksmithing could be a relatively well-paid occupation. Lotoy purchased his iron in Iboko, buying imported rebar at 180 Z per meter, smooth bars of slightly higher quality imported iron at 250 Z per meter. Approximately thirteen arrow blades could be made from one meter of bar, each one selling for 100 Z, leaving a net intake after materials cost of 1050 to 1120 Z. Although thirteen arrow blades could be made in one day, it is unlikely that they would all be sold immediately. Moreover, the forms of payment were variable, subject to fluctuations in the supply of cash in the local economy. Smiths' customers often exchanged goods or produce—charcoal or game from Twa hunters, for example—in order to acquire iron products. Smiths' incomes stood up well in comparison with the 400 Z per day considered good wages in 1989,[51] and this evidence is consistent with testimonies that in the remembered past it was the *smiths* and not the smelters who became rich because they made all the finished products.[52]

Only the master blacksmith, Lotoy, was able to make more complex and technically demanding products, the elaborate knives which involved the greatest control and mastery of a variety of smithing techniques and sequences. I commissioned Lotoy to manufacture one particular type in order to resolve some questions I had about just how its forms were achieved. Making this knife, known as *ikákáláká*, took about four days of continuous work, and throughout the process Lotoy was constantly commenting to others on how he was planning to arrange the ornamentations as he went along. Periodically, others would make suggestions, but clearly Lotoy took this opportunity to display his singularly superior prowess at the forge. He gave special care to the splitting and forming of spikes and tendrils that were regularly and symmetrically placed along the axis of the blade. When its form was completed, he blackened the blade's surface with sugar cane and selectively polished it along the edges, a contrast that heightened the bright, silvery gleam of the iron. It was admired throughout the town as a work of art.

Such knives could be a significant source of profit for blacksmiths, though in aggregate numbers their production was undoubtedly surpassed by that of arrow blades and lances. Even in the very depressed economy of 1989 this was so. The *ikákáláká* knife sold for 2500 Z, costing 250 Z in materials and about four days of labor, netting Lotoy 2250 Z or about 560 Z per day. Working knives, no longer made locally, would have sold for about 600 Z, and, while requiring the same material cost but taking only one day of labor, would have netted Lotoy 350 Z. These current pricing data indicate little regarding what prices were in the nineteenth or early twentieth century. However, it is

Fig. 7-1: *Ngolo* sword. Iron blade, with incised, stamped and grooved surface patterning, brass wire and brass tacks on wooden handle. Blade, 37.7 cm long. Courtesy of the Royal Ontario Museum, Toronto, Canada. 948.2.9. Collected by T. Hope Morgan before 1910. [Photograph by the author]

clear that a relatively significant cash return could be gained in one lump sum for the sale of elaborate knives.

Four of the five types of large, elaborate knives (*lokulá lombole, ingóndá, ikákáláká, ibaka,* and *bompata*) displayed in Lopanzo had been made by Lopanzo's past master smiths, at least three of them by smiths in the nineteenth century. The techniques for making them survived to the present, but the reasons for their invention and some understanding of the important social messages they have conveyed must be sought in the precolonial period. Added to the economic value of at least some of these knives was their considerable social value, since they served not only as weapons but also in some cases as insignia for local and territorial leaders. Their histories reveal once again the important social networks formed by blacksmiths through their products and the ways some of these products were worn and wielded as markers of social prominence.

LOKULÁ LOMBOLE

Of the five luxury knives used and displayed in Lopanzo, the *lokulá lombole* was the only one that the master blacksmith, Lotoy Lobanga, was not able to make. It was an elaborate, single-edged sword, characterized by a wide curve of the blade and parallel incised lines along the blunt, scalloped side of each of the blade's faces.[53] It has sometimes been mistakenly described as a "serpe," or billhook; however, billhooks have their cutting edge along the concave side of a blade's curve, while this sword had its cutting edge along the convex edge. Its design was an innovation based on a prototype sword that had two curved edges, this new one created simply by dividing the older sword design in half, longitudinally.[54]

Missionaries and colonial officials from the 1880s onward noticed prominent men carrying this impressive single-edged sword and began referring to it as a "Bangala execution knife." The first time an illustration of it was published, this and other luxury knives were described very generally as metal goods that were traded along the middle Zaire.[55] Soon, however, descriptions of this sword were refashioned to serve a colonial myth of tribal savagery. A colonial officer at the Equatorville Station identified it as a "knife for sacrifice (m'boulou) of Ba-Ngala," and included it in his illustrated description of male slaves being executed at the funeral of a chief in 1883.[56] Association of the sword with the so-called Bangala has been persistent, even though the execution described did not take place near the Bangala Station, but was at the funeral of the grand chief of Wangata, further downriver near Equatorville. And at least one other type of sword was illustrated being used in an execution.[57] Added to the numerous hearsay reports by colonial officers and missionaries that human sacrifices were carried out at the funerals of important chiefs, these two illustrated eyewitness accounts became useful pro-colonial propaganda during the time when Leopold's regime was about to be transferred to Belgian state authority. The original 1883 funeral account was repeated, this time with a photograph of the sword labeled "Bangala Execution Knife," in a 1907 article for the *Moniteur Colonial*.[58]

These sensational descriptions of how such swords were used defined them in an almost irrevocable way, though there were additional uses for them that were recorded by others. Prominent men along the Zaire River between Malebo Pool and Upoto owned and carried this type of sword, called *ngwolo* or *ngolo*, in public on ceremonial occasions (Fig. 7-1).[59] Indeed, it appears that such luxury goods were most frequently kept or owned by village or territorial leaders and were used by them to display the current legitimacy of their authority. At the same time, however, the sword was capable of serving the more practical purposes of a weapon, and when executions were carried out at the public funeral for a leader (though it is still unclear exactly who the victims were or why they were killed), it is understandable that an elaborate knife of authority would be used. But to call it simply an execution knife and to associate it with a single ethnic group is woefully inaccurate.

It was a sword often displayed by men in the central basin who held the title of *nkúmú* in the nineteenth and early twentieth centuries.[60] *Nkúmú* is an institution estimated to have a history going back perhaps as far as seven hundred years in the region around Lake Mayi Ndombe. Vansina provides a concise description of the institution and its variants as they developed in different social environments.[61] It was primarily a titled position acquired by individual men through their own achievements, not by inheritance, and in many areas led to the creation of chiefdoms or kingdoms organized around the *nkúmú* as recognized leader. Descriptions of his authority combine attributes of judge, diviner, and political administrator legitimized by both sacred and secular sanctions.[62] Hence the photographs published that show *nkúmú* title holders displaying knives identified as "execution knives" seriously distort this important precolonial institution.

It is not clear exactly where or when the single-edged *ngolo* was invented, but it may have been inspired by the scimitar. *Ngolo* and several other types of curved sword were manufactured and traded widely in the northern Zaire basin in the second half of the nineteenth century, perhaps owing to the power and prestige at that time of Muslim overlords and soldiers.[63] The *ngolo* sword's double-edged prototype was identified as a product of numerous smithies along the Ngiri watershed in the 1880s, but the single-edged version was then still the monopoly of ironworkers at several distinct centers. Only one of the late nineteenth-/early twentieth-century blacksmiths who worked in two productive smithing villages along the Ngiri River had known how to make a *ngolo*, and he had traveled one hundred km for the retraining. With the help of river merchant intermediaries, he arranged to work with a Monzombo blacksmith in the town of Dongo on the Ubangi River, where he learned the skills for manufacturing this new type of sword.[64] Blacksmiths from Ngombe towns around Mampoko, near the Lulonga River, had also acquired the specialized skills for making it, at least by the turn of the century.[65] These smithies were another one hundred km away from the center on the Ngiri.

Some smiths today are still able to make this special sword, though demand for them has fallen. In Lopanzo on the first day in August, 1989, a *nkúmú* candidate from Bongila passed through on his preliminary rounds, his supporters hoping he would fill the power vacuum left by the *de facto* collapse of Mobutu's administration. He carried a lance along with the curved sword, each with red and white pigmentation on the surface of the blade. I was told that the knife was the knife of a *nkúmú*, and that a former *nkúmú* title holder of Lopanzo had had one that was passed down to his son, though the position of *nkúmú* was not. In Lopanzo, the sword was called *lokulá lombole*, or knife of the Mbole. This was because it was made by Mbole blacksmiths of the town of Monkoto on the Luilaka River, southeast of Lopanzo in the central basin. Master blacksmith Lotoy could not make one, and neither could any of the other former Lopanzo smiths. Someone who entered into candidacy for the position of *nkúmú*, or anyone else who wished to purchase the sword, would have had to travel to Monkoto to have it made, a distance of about 150 km.

If precolonial political and judicial institutions had not been weakened by colonial rule, perhaps the demand for these swords would have continued and perhaps then a smith from Lopanzo would have invested in the retraining for making it, traveling to Monkoto and working for a time with smiths there. Exactly when and by what means the Mbole blacksmiths themselves had initiated and arranged for their own retraining and especially whether they traveled as far as the smithies around Mampoko (350 to 400 km away) is not known. What is certain is that such retraining was necessary, for technically complex products like the *ngolo* sword could not be replicated simply by copying, but required mastery of specific skills and production sequences for achieving its form. Blacksmiths did invest in such retraining, much more so in the past, traveling distances and crossing ethnic and language boundaries to upgrade their level of expertise.

INGÓNDÁ

The most historically significant luxury knives made in Lopanzo were a type of double-edged sword called *ingóndá* (Fig. 7-2).[66] It was characterized by a splayed, fanlike form at the tip end of the blade, and blacksmiths often finished its surface by blackening and selective polishing. The midrib was almost always a symmetrical one, branching out into two or more ribs at the splayed end. It had beveled cutting edges on both sides of the blade and sufficient mass to have been used as an effective cutting or chopping weapon.[67] This particular blade form with its splayed end was probably derived from lance blades made to perform as cutting missiles, used especially for hunting elephant and other large game.[68]

Ingóndá knives were certainly effective also as close-range weapons. That they were carried by ivory and slave traders from the upper Zaire in the second half of the nineteenth century attests to their practical uses as defensive and offensive weapons against other human beings.[69] Present day inhabitants of Lopanzo repeatedly stressed the fearsomeness and efficacy of *ingóndá* as a personal weapon. Three examples of the sword were in the possession of the family of Georges Boleteni, who demonstrated to me how they would have been carried for protection while traveling, or used as a weapon in war or in the hunt. Two of them had two midribs along the length of the blade, while the other had only one midrib. All were considered to have been the work of the smith Iponga, who died in the first decade of this century.[70]

But beyond what people were willing to state directly in words, it was clearly demonstrated to me that *ingóndá* was more than a weapon, that it bore a number of significant social values as well. The three main houses of Lopanzo each kept collective treasuries which included *ingóndá* swords, in each case the sword being associated with a renowned blacksmith who had belonged to that house. The current elder was the privileged keeper of the treasury, and in naming past keepers of the sword, the house history could be recounted. In this way, treasures such as the *ingóndá* served as mnemonic devices for reciting the names of past leaders.

The treasury swords were also masterworks that could serve as teaching or replication models for blacksmiths, and they were permanent records of the ways blacksmiths had further refined their products and techniques. Ultimately, they revealed the differing levels of skill among masters. I was able to examine two of the three house swords, and some other examples kept by families in the town. It became clear to me that there was a significant but subtle hierarchy of *ingóndá* swords based primarily on a single technical feature, the number of midribs. The variant that was most highly valued and most difficult to make had three midribs, hence it was also called *ebale*, a term of praise.[71]

Collectively held house swords that I examined were from the Ipumba and Nzelika treasuries. The one belonging to Ipumba was referred to as the work of the master smelter-smith Bongongo Lompulu, whose peak productive period was c. 1880 to 1910. It had two midribs, and was also split at the midpoint of each cutting edge. Lotoy claimed that to make such a sword took three or four days.[72] The Nzelika sword was an elegantly shaped *ebale* type, with three midribs. It was reputedly made by Nsabonzo Engala, the master smelter-smith referred to earlier in this chapter, who also worked from about 1880 to 1910.[73] Although the sword was very much corroded, it was

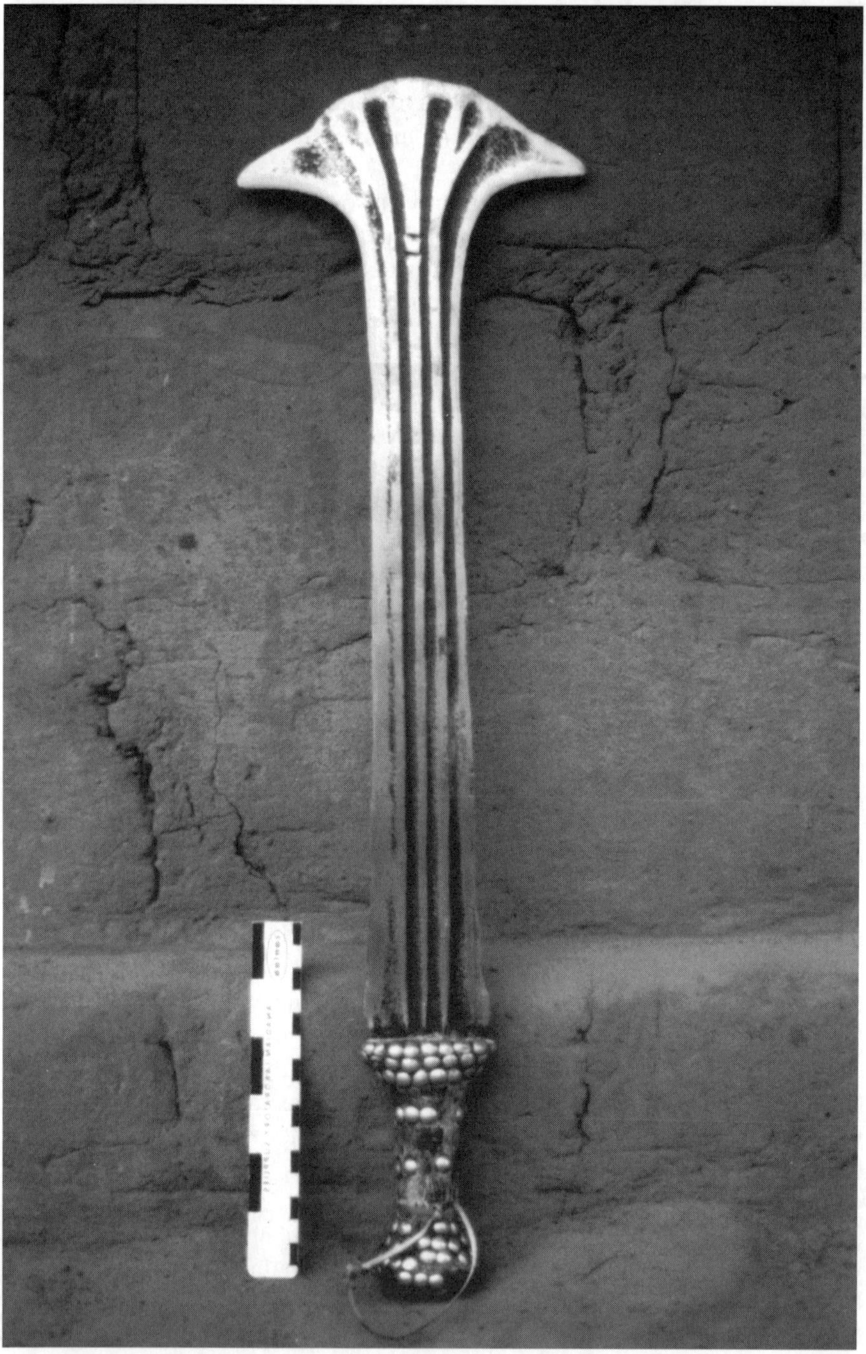

Fig. 7-2: *Ingóndá* knife with three midribs (*ebale* type). Iron blade, blackened and polished. Reportedly made by Nyekuli Bana, master smelter-smith, in Lopanzo during the 1930s. [Photograph by the author]

still possible to recognize the high level of control and skill that had been used to draw out such a form with three parallel and symmetrical midribs. Here it should also be noted that the *ingóndá* swords in museum collections were almost always those having one or two midribs; the *ebale* types must have been fewer in number, more valuable, and not as alienable as the less elaborate types.[74] These treasured swords may have been indeed the masterworks of these well-known smelter-smiths, or they could have been even older, the teaching models that had belonged to their own masters. Whatever the absolute ages of the objects were, their designs represent a particular constellation of skills and techniques that blacksmiths had developed at least one generation earlier.

Another *ingóndá* that I examined helped me to identify other important social uses for the sword. It was an *ebale* type in the possession of Nsabonzo Bilanga, who was the head of Lopanzo in 1989. It reputedly had been made by Nyekuli Bana, the master smelter-smith of the Nzelika house who had recently died. He had passed it on to his son, Nsabonzo Bilanga, who displayed it on important public occasions as the "sword of Lopanzo." That it was so used explains why it had been kept in such carefully oiled and polished condition, with no trace of rust at all. Nsabonzo claimed that he also used it for travelling, and especially for showing and dancing at a *bonkunda*, the funeral of a house chief. Above all, it was claimed by him and others that the sword had been made "for the pride of men,"[75] but of course only certain men. Lotoy stated that *ingóndá* would be carried by the "chef de famille" as a sign of superiority at a *bonkunda* ceremony, that it was never carried by women, and could only be carried by Twa men who had special permission. The spokesman for the Nzelika house also claimed that *ingóndá* was a sign of pride and superiority, carried by the one who kept it, the house elder, especially to *bonkunda* ceremonies. He was insistent that one never used such a sword for work, that women could never use it, and that only wealthy Twa men with permission could carry one.

These recently spoken testimonies are supported by archival photographs from earlier in this century in the central basin.[76] They show formally arranged presentations of wealth, prestige, and authority in places where precolonial institutions continued to persist along with some of their significant paraphernalia. One such photograph is of a leader sitting on his official stool, surrounded by his wives all bedecked with cast copper alloy necklaces. Standing behind him were three men with lances and swords, one of them an *ingóndá*, and behind them all was an impressive backdrop of leopard skins. The leader appears to have been a *nkúmú* title holder.[77] Another photograph, taken in the 1930s, shows a chief and elders and notables from villages around Bolia reenacting the history of the Lia migrations. Men carrying clusters of lances in the left hand and swords in the right, accompanied the chief who was himself carried in a seat representing a canoe. Four of the five visible swords were *ingóndá*, all with blackened surfaces and polished edges.[78] In a photograph of chiefs or *nkúmú* title holders taken north of Lake Mayi Ndombe, they were wearing the elaborate, tiered caps of leadership, and several held swords, two of them being *ingóndá*.[79] Still other photographs show notables, titleholders, and diviners carrying this type of sword.[80] In short, there do not seem to have been any stringent rules regarding which individuals among elite men could or could not own or display an *ingóndá*—just that women could not.

IKÁKÁLÁKÁ OR *BYONGI*

Blacksmiths may have developed this type of knife as a more highly ornamented extension of the *ingóndá* sword.[81] Unlike the *ngolo* and *ingóndá*, there was no consistent, standard form for this product, and it was characterized instead by the variety of elaborate forms it could take. Blacksmiths utilized ornamental cutting and splitting techniques to produce it, thereby exhibiting their virtuosity in controlling and balancing visual elements along the vertical axis of the blade.

The *ikákáláká* knife appears to have been the invention of blacksmiths south of the lower Kasai, sometime in the nineteenth century. There it was a knife split at the tip into two points, produced in two sizes: a small, lightweight knife; and a heavier, more massive version that was associated with war, the position of chief, and judicial authority.[82] By the late nineteenth century, the smaller version of it was being manufactured in smithies around Lake Mayi Ndombe and Lake Tumba, where it was a prestige knife for chiefs and other notables. Eventually it was used and made over much of the central basin, displayed publicly by titled men at festivities and on ceremonial occasions.[83]

But unlike the other ornamental knives *ngolo* and *ingóndá*, the *ikákáláká* was sometimes displayed by prominent women as well as men. In most of these instances, the women were holders of titles bestowed on them not for their individual achievements but based on their position as the principal wife of a powerful man.[84] A woman who held the title of *bolúmbú*, for example, was awarded it by either her own husband or, in some locales, by a council of male elders. She was then designated as holding a special social position, having legitimized authority to intervene in and resolve disputes. That position was publicly recognizable by her right to wear fur, attire reserved exclusively for men, and her display of other prestige goods such as the ornamental knife *ikákáláká*.[85] Similarly, the title of *nsónó* was given to the woman selected to be the principal "wife" of the *nkúmú*. Officiating at her selection was a specialist who tested candidates for divine sanction. The *nsónó* was to be the "wife" closest to the *nkúmú*, taking part in some of the ritual aspects of his investiture and sometimes representing him in his absence.[86]

All of these contexts for displaying this knife were confirmed in interviews in Lopanzo. Men and women alike described the *ikákáláká* not as an implement that could have been used in war or for defense while traveling, but as an ornamental knife, one that in recent years was often sold to Europeans.[87] But local uses for it had not entirely disappeared and were not insignificant. Either men or women could dance with it at *bobongo* festivities,[88] and there were no particular regulations concerning who could keep or be given an *ikákáláká* knife. Women even made the claim that it could be passed down along the female line as women's property. Such would be the case perhaps when women acquired the knife as part of diviners' prescribed treatments for certain forms of mental or emotional disorders. Dance therapies were often performed with an *ikákáláká* knife held in the right hand, the blade pigmented with white on one half, red on the other.[89] In other cases, women would display a knife that was their husband's property, as in the ceremony called *walé*, marking the end of a woman's period of seclusion following the birth of her first child. An important part of the ceremony was the woman herself, seated, formally exhibiting her husband's wealth in copper ornaments and other paraphernalia, including the *ikákáláká* knife.[90] This was one luxury knife that was not a male monopoly, though it was mainly exceptional women of title and high status who owned or displayed it, a privilege bestowed upon them by men.

IBAKA

Unlike the three swords and knives already discussed, all of which were cutting or chopping blades, this blacksmith's product was a cut-and-thrust blade, a personal dagger with a rounded point. At its lower center, near the handle, was an area of surface patterning in parallel incised lines. Another distinctive feature of this knife was that it was carried by men in a carved wooden sheath under the arm, both sheath and handle often embellished with brass tacks. I was told in Lopanzo that the knife was called *ibaka*, though another name for it was *empúte*.[91]

It, too, has a history that goes back beyond the middle of the nineteenth century. Blacksmiths in the central basin manufactured a very similar blade form, though without the incised lines in the center, and this plain blade was still being widely traded from the middle Tshuapa and Maringa Rivers down to the Zaire in the 1880s.[92] It is likely that this was an earlier product, the prototype on which the more elaborate *ibaka* knife was based. Its only adornment was a blackened surface with branching midrib pattern, features shared also by the *ingóndá* sword of the central basin. Blacksmiths along the middle Zaire, who invested in engraving tools and learned the skills of incising linear patterns on metal surfaces, invented the *ibaka*, a new, more luxurious product, by applying this ornamental technique to a standard type of blade that was already well known.[93]

In doing so, these blacksmiths added yet another product to the prestige market, luxury goods that were purchased by ambitious, upwardly mobile patriarchs in the nineteenth century. In some of the earliest reports by Europeans, they noted that the *ibaka* knife was considered part of the essential equipage of freemen, notables, and chiefs in and around the Equatorville station, south from there to Bolobo, and inland.[94] It was owned by a wider cross section of the male elite than were other examples of prestige goods, and in many cases it was worn in its sheath under the arm while one of the other types of sword was more prominently displayed for show. That it had more to do with social status than political authority was confirmed in Lopanzo. There, in the past, formal male dress was described as having included a lance, in the case of an elder, and also an *ibaka*, for the eldest in the village. It was more directly and consistently carried as a personal weapon and only as a weapon, and was owned by individuals, never held collectively as the property of a chiefdom, house, or family group.[95] Moreover, unlike the various types of swords, there were no references made by anyone to prohibitions against certain individuals touching or handling this knife.

BOMPATA

This type of knife was less prestigious and not as well established in Lopanzo as the others. It was characterized by a long, thin, cut-and-thrust blade presumably derived from a lance blade prototype.[96] Most examples were very light and thin, and so the *bompata* knife would not have been an effective weapon for personal defense. Also, relative to the other swords and knives, it was very simple in construction, and so manufacturing it required about two-thirds to three-quarters less time. Blacksmiths did not have to be particularly well skilled to make it, so it could have been easily replicated from a trade blade, and this was probably how it came to be manufactured at forges in the Lake Mayi Ndombe area. More of them had circulated as currency blades in the late nineteenth and early twentieth centuries in areas along the upper Lukenie River further to the east and south of Lopanzo.[97] Being visually unremarkable, though, meant it was not a particularly coveted object of prestige among the wealthiest men in society.

IRONWORKERS AND SOCIAL MOBILITY IN LOPANZO

To sum up, this chapter brings into sharper focus the range and depth of the impact ironworking and ironworkers could have on a local community, and how the multiple uses and values of iron could shape aspects of economic, social, and political life in nineteenth-century central Africa. The case of Lopanzo shows that in the past, smelter-smiths were geographically mobile, and so could prospect and choose sites for new settlements near ore deposits, assuring themselves of a continuous supply of iron. Once again, the evidence suggests that the value of ore deposits did not lead to absolute monopoly control over them. Claims over deposits were most valuable not when land chiefs restricted access to and exploitation of ore, but when they merely regulated it. Insiders and outsiders were able to gain access to ore and iron blooms through negotiation and division of the mining and smelting proceeds, and it was master smelter-smiths who, as managers, made the key decisions granting individuals access to the resource area and working processes.

Also apparent here are the complexities and inequities that existed in terms of access to skills. Smelters and smiths accumulated considerable social and economic power as a group by withholding the transmission of valuable skills to women, Twa, and to most men outside their own houses. They did so dramatically, with the verbal and visual flourishes of ritual that served also to amplify the importance of the work and the workers. But ironworkers did not confine their occupation completely within their family or lineage structures, a strategy that surely enhanced, rather than weakened, their prominence. Master smelters and smiths enjoyed tremendous social leverage by exercising their discretionary power as owners of skill, information, and, by extension, the final product. That, added to their patriarchal authority over family members, slaves, and clients, allowed them to shape the labor process to suit their own individual and changing needs. When they permitted select non-kin assistants and trainees to join their workshops, technical knowledge spread laterally, permeating kinship, village, and ethnic boundaries over time. This case study of Lopanzo ironworking illustrates more clearly how this could have been done, and it can be safely inferred that master smelters and smiths would have been the ones to initiate these transfers.

In promoting their work and their own reputations, ironworkers rose to prominence and remodeled the populations of their communities. They generated wealth by winning metal from worthless iron ore and by making value-added products that could be sold to everyone at every social level. Then, with their own considerable wealth in iron, they were well positioned to invest in the women, slaves, and other human capital that supported and created their large houses. They made other investments, in retraining and retooling, which helped them become well known outside their towns and villages, lent dynamism to their careers, and brought them into contact with ironworkers at regional centers and with powerful patrons in other communities. The social stratification they had a hand in generating was in turn demarcated visually by their very own products, in the hierarchies of prestige goods that blacksmiths invented and mastered.

As powerful men and transformers of society, successful ironworkers created names for themselves that were long remembered. Their successes could hardly be overlooked. The iron they made brought women to their houses and towns, and drew clients and customers to their workshops. One could witness success also in the quality of their work, the best of it admired by all in public ceremonial displays by political leaders, judges, and notables, including some select women. After their deaths, the most renowned ironworkers attained the status of revered ancestors. Nsabonzo Engala and Bongongo Lompulu, for example, were hailed by their descendants, who called out their names over and over in song, and who cared for and protected their masterworks as communally held treasures.[98]

Prestige goods, many of them made by blacksmiths, are a window onto the visual representations of wealth and social hierarchies in nineteenth-century central Africa. A formal display made in Lopanzo by its head, Nsabonzo Bilanga, and the designated master smelter, Isolo, conveyed simultaneously several layers of possible messages.[99] It was above all an elaborate historical statement, a tribute to the achievements of the most recent master smelter-smith, Bana. Walking in procession, Isolo wore the furs and other paraphernalia that had identified Bana as a master and member of the male elite, while Nsabonzo carried his precious sword upright, orienting it deliberately to present its gleaming face, not its edge. So displayed it was an object of beauty, not a weapon. In my view, the two men together were celebrating and offering visual proofs of Bana's legacy, his past position of master smelter, his insignia, and his masterpiece, an *ebale* type of sword. This formal presentation thus served as a political statement for the present, an affirmation of Nsabonzo's legitimacy as the current mayor. Proving Bana's position and legacy shored up his own immediate position and that of his house. At the same time, such a ceremonious display insured that Bana's achievements would be admired, remembered, and carried forward in time by the next generation.

NOTES

[1] Korse, "La Forge," p. 31.

[2] Manfred Eggert, "Aspects de l'Ethnohistoire Mongo: Une vue d'ensemble sur les Populations de la Rivière Ruki," *Annales Aequatoria* 1 (1980), pp. 162–63.

[3] Eggert, "Aspects," p. 161 fn 4. I am deeply indebted to Kanimba Misago for sharing with me the genealogies he collected and verified in Lopanzo. See his IMNZ report C/ADR/12.

[4] H. Rombauts, "Ekonda e Mputela," *Aequatoria* 9, 4 (1946), p. 122.

[5] For a brief description of a Nkundo forge, see A. de Rop, "Le Forgeron Nkundo," *Annales de Notre-Dame du Sacre Coeur* (February 1956).

[6] Kanimba Misago, IMNZ, report C/ADR/12, p. 6.

[7] H. Rombauts, "Les Ekonda," *Aequatoria* 8, 4 (1945), p. 125.

[8] Hulstaert, "Aux Origines de Mbandaka," p. 120; Eggert, "Aspects," pp. 151–52. These foundry remnants, excavated by Eggert, were discussed earlier in Chapter 3. The furnace was a bowl type with slag-tapping tunnel and slag pit, dating to 1680–1800 AD, a period after the estimated time of the influx of Nkundo populations. See Eggert, "The Current State of Archaeological Research," p. 41.

[9] Rombauts, "Les Ekonda," p. 125.

[10] Célis, "Fondeurs et Forgerons Ekonda," p. 112.

[11] Célis referred to twelve forges in Lopanzo in 1974. It is not clear whether these were active or not. Ibid., p. 111. In 1989, I viewed many hammers and anvils kept by house leaders and never offered for sale, though I made no formal count of just how many there were.

[12] In Célis's account of the 1974 smelt, he states that the smelters claimed their ancestors came from the Tshuapa area 200 km east. Ibid., p. 110. This contradicts the 1989 account of their origins by the Nzelika house, an account that was corroborated by genealogies collected by Kanimba Misago in neighboring villages.

[13] Kriger, field notes, 17 July 1989. The tree *boséngа* is *Pychanhus kombo* Warb., according to G. Hulstaert, *Notes de Botanique Mongo* (Brussels: ARSOM, 1966). I thank Dr. Eugenia Herbert for sharing with me her notes on medicinal and other plants associated with the smelt in Lopanzo.

[14] Kriger, field notes, 26 July 1989, 17 July 1989.

[15] They were not as skilled, however, as were smelters of the past in selecting the highest grade ores. See Ackerman et al., "A Study of Iron Smelting at Lopanzo."

[16] Kriger, field notes, 19 July 1989; Kanimba, IMNZ report, p. 9; C. Kriger, Log of Lopanzo Iron Smelt, in idem, *Ironworking*, pp. 427–28; and Ackerman, et al., "A Study of Iron Smelting at Lopanzo."

[17] Interview, J. E. Rehder, 27 September 1989.

[18] Laboratory analyses of ore samples, both roasted and not, show that the ore was insufficiently broken up and prepared for the smelt. See Ackerman, et al., "A Study of Iron Smelting at Lopanzo."

[19] Kriger, Log of Lopanzo Iron Smelt, in Idem, *Ironworking* and Ackerman, et al., "A Study of Iron Smelting in Lopanzo"; Célis, "Fondeurs et Forgerons Ekonda," pp. 118–19.

[20] See Ackerman, et al., "A Study of Iron Smelting in Lopanzo."

[21] Kanimba, IMNZ report, p. 9

[22] Interview, J. E. Rehder, 28 December 1990; and Tylecote, Austin, and Wraith, "The Mechanism of the Bloomery Process," p. 363.

[23] Interview, J. E. Rehder, 28 December 1990.

[24] I assisted J. E. Rehder in two laboratory iron smelts carried out in May of 1988 for the 1988 Symposium on Archaeometry, held at the University of Toronto. The first, which yielded a bloom of 1.1–1.2 percent carbon iron, was successful; the second only yielded some iron fragments.

[25] See especially Herbert, *Iron, Gender, and Power.*.

[26] Interview, Ndongo Bokoti, 22 July 1989.

[27] Kriger, field notes, 17–19 July 1989. Townsfolk all emphasized that the three North American women on the research team "were not women," that is, we were recognized as women biologically but not socially.

[28] See Guyer, "Indigenous Currencies"; idem, *Money Matters;* and Caroline Bledsoe and Gilles Pison, eds. *Nuptiality in Sub-Saharan Africa* (Oxford: Clarendon Press, 1994).

[29] G. Hulstaert, *Dictionnaire Lomongo-Français* (Tervuren: Musée Royal du Congo Belge, 1957), Vol. II, p. 1155. For a reconstitution of a chief's marriage transaction show-

ing the public transfer of old copper currency from the early colonial period, see the 1951 photograph taken by Olga Boone in Ngongo-Iyembe. MRAC, Department of Ethnology, EPH 2.296.

[30] R. Bebing, "Réponse au questionnaire sur le droit indigène" (1931) Territoire Bikoro 2, District Equateur, Province Equateur, Documents, Department of Ethnology, MRAC, pp. 9–10; personal communication, Boilo Mbulo and Kanimba Misago, 20 July 1989; G. Hulstaert, "Notes sur le Mariage des Ekonda," *Aequatoria* 11 (Dec. 1938); Vansina, *Paths*, pp. 107–109.

[31] Pieters, "Rapport sur l'Administration, Coutumes, et Moeurs de la Tribu Ekonda" (1925) Territoire Bikoro 8, District Equateur, Province Equateur, Documents, Department of Ethnology, MRAC, p. 2; Bebing, pp. 9–10; personal communication, Kanimba Misago, 20 July 1989; group interview, elder women, Iyenge, 28 July 1989.

[32] The introduction of brass wire currency was tied to forced labor, especially the collection of red rubber for export and the administration of colonial posts. Official figures show that the upper Zaire was just behind the lower Zaire area in the amounts of rubber supplied in 1892; by 1897 the upper Zaire was supplying almost 15 times the amount steadily coming out of the lower Zaire. *Bulletin Officiel de l'Etat Indépendant du Congo* No. 4 (1893), pp. 53–61 and No. 3 (1898), pp. 56–65. For a general discussion of the introduction of brass currencies see Harms, *River of Wealth, River of Sorrow*, p. 92.

[33] These bars were 100 x 5 x 1 cm. Kriger, field notes and photograph, Lopanzo, 1989; group interview, male elders, Lopanzo, 24 July 1989; group interview, women elders, Iyenge, 28 July 1989; group interview, women elders, Lopanzo, 21 July 1989.

[34] De Rode, "Note sur la fonte du fer," *Aequatoria* 3, 4 (1940).

[35] Interview, Elombola Boponde, 29 July 1989; Kanimba IMNZ Report, p. 15.

[36] Bebing, p. 8; Mumbanza mwa Bawele, "Les Forgerons de la Ngiri," pp. 124, 132 fn 16.

[37] As far as is known, there were no Twa smiths in the area immediately around Lopanzo. Group interview, male elders, 18 July 1989 and 24 July 1989. Regarding Twa smiths in Rwanda, see G. Célis and Emmanuel Nzikobanyanka, *La Métallurgie Traditionelle au Burundi* (Tervuren: MRAC, 1976).

[38] Group interview, 24 July 1989. See also Célis, "Fondeurs et Forgerons Ekonda," p. 111.

[39] Interview, Elombola Boponde, 29 July 1989.

[40] Interview, Lotoy Lobanga, 25 July 1989.

[41] One set of anvil and hammer were inherited and kept by Nsabonzo Bilanga, the *capita* (mayor) of Lopanzo, although he did not use them. Interview, Elombola Boponde, 29 July 1989. See Célis, "Fondeurs et Forgerons Ekonda," p. 111, for an anecdotal account of ceremonies surrounding the completion of apprenticeship and the formal presentation and purchase of tools, and pp. 122–23 for ceremonies surrounding the manufacture of anvils and hammers. Four-drum bellows were needed to generate high enough temperatures in the forge for heat welding masses of iron in manufacture of these tools.

[42] Interviews, Lotoy Lobanga, 25 and 27 July 1989.

[43] Célis, "Fondeurs et Forgerons Ekonda," p. 133.

[44] Interviews, Lotoy Lobanga, 25 and 27 July 1989.

[45] Personal observation of the work of Nsatale Iseke, 25 July 1989; Interview, Lotoy Lobanga, 29 July 1989. However, I have photographs of the microstructure of part of a hoe blade collected in Lopanzo, and it shows a welding joint for which some flux was probably used (inference based on the absence of FeO). Interview, J. E. Rehder, 28 December 1990. See Kriger, *Ironworking*, p. 429.

[46] Personal observation, 26 and 27 July 1989; Interview, Lotoy Lobanga, 29 July 1989.

[47] Interview, Lotoy Lobanga, 26 July 1989. See Hulstaert, *Dictionnaire Lomongo-Français*, Vol. I, p. 214.

[48] The sample does not show evidence of martensite, which is produced by quenching. This would have been a low carbon martensite. Interview, J. E. Rehder, 28 December 1990. See Kriger, *Ironworking*, p. 430.

[49] Interview, Lotoy Lobanga, 27 July 1989.

[50] Interview, Lotoy Lobanga, 25 July 1989.

[51] Interview, Lotoy Lobanga, 27 July 1989.

[52] Group interview, 24 July 1989.

[53] Kriger, museum notes, technical data from the 115 examples I surveyed in the Department of Ethnology, MRAC, Tervuren.

[54] See my proposed seriation, showing prototype swords and the successive innovations spawned from them in Kriger, *Ironworking*, pp. 249–53.

[55] Johnston, *The River Congo*, p. 434.

[56] Coquilhat, *Sur le Haut-Congo*, pp. 241, 168–74.

[57] Glave, who was stationed at Lukolela, Bolobo, and Equatorville from 1883 to 1889, made a drawing of another sword being used to decapitate a prisoner or slave, calling it an executioner's sword. Glave, *Six Years*, pp. 124–25.

[58] *Le Congo. Moniteur Colonial* 4, 135 (28 April 1907), pp. 132–33.

[59] A chief in the environs of the Bangala Station was photographed in 1896 surrounded by his wives and holding the knife in a position of formal display. G. Van der Kerken, *l'Ethnie Mongo*, 2 Vols. (Brussels: ARSOM, 1944), Vol. 1, Plate LIII, No. 231. Thonner described it as a sword for parade and for justice. Thonner, *Dans la Grande Forêt*, Plate 35. The lexical term for it was known in the C languages of Bobangi, Ngombe, and Lomongo, and probably others.

[60] See Jean Leyder, "Notes sur le couteau 'ngwolo' des Ngombe de la Lulonga (Congo Belge)," *2e congrès National des Sciences, comptes Rendus (Brussels)* II (1935); André Scohy, "Notes sur la perle 'nkangi' ou 'nkange'" written at Gombe, 1 June 1939, Territoire Bikoro 2, District Equateur, Province Equateur, Documents, Department of Ethnology, MRAC, p. 4; B. Blackmun and J. Hautelet, *Blades of Beauty and Death* (San Diego: Mesa College Art Gallery, 1990), Plate 8; MRAC, Department of Ethnology, EPH 14546, EPH 12.217; Célis, "Fondeurs et Forgerons Ekonda," p. 130.

[61] Vansina, *Paths*, pp. 120–27. See also idem, *Introduction à l'Ethnographie du Congo* (Kinshasa: Université Lovanium, 1966), p. 80; G. Wauters, "L'Institution du 'nkum,'" *Annales des Missionaires du Sacre Coeur* (1937); and J. B. Stas, "Les Nkumu chez les Ntomba de Bikoro," *Aequatoria* 2, 10–11 (Oct.–Nov. 1939).

[62] See my brief discussion of *nkúmú* costume in C. Kriger, "The Knife and the Hoe: Gender, Metals, and Sources of Power in the Middle Zaire Basin, late nineteenth and early twentieth centuries," paper presented at a conference at Oxford on Technology and Gender (May 1993), to be published in a volume edited by Ian Fowler.

[63] Some of these swords only resembled scimitars superficially, for (in contrast to the ngolo) they had their cutting edge on the concave side of the blade, not on the convex side, which is the distinguishing feature of a scimitar. For examples of such pseudoscimitars, see Vansina, *Paths*, pp. 107–78.

[64] Mumbanza mwa Bawele, "Les forgerons de la Ngiri," p. 123.

[65] Leyder, "Notes sur le Couteau 'Ngwolo'," pp. 941–43. Smiths claimed the sword was made with imported iron in the 1930s, but had previously been made with locally smelted iron. This comment suggests the Mampoko smiths were manufacturing it at least in the early years of the twentieth century, probably well before.

[66] Interview, Lotoy Lobanga, 25 July 1989; see also Hulstaert, *Dictionnaire Lomongo-Français*, Vol. I, p. 836.

[67] Kriger, museum notes, technical data based on the 107 examples I surveyed in the Department of Ethnology, MRAC, Tervuren; and nine examples in the collections of Centre Aequatoria, Mbandaka, Zaire (collected in the 1980s). Eight of the latter show differences in form which appear to be attributable to differences in area workshops. These eight were blunt at the tip, without the rounded curve tapering into sharp points. Hulstaert also shows a blunt-tipped variant in his Lomongo dictionary, calling it *efefa*. Hulstaert, *Dictionnaire Lomongo-Français,* Vol. I, p. 488.

[68] One such hatchetlike lance blade was already discussed in Chapter 5, since it was a blade currency in the northern Zaire basin and Uele watershed. A smaller variant circulated in the trade networks around Malebo Pool. It was called *etoála*, but ceased being made in the second half of the twentieth century. Interview, Lotoy Lobanga, 25 and 27 July 1989. Both types probably were traded into the central basin in the nineteenth century.

[69] Jeannest noted them as part of the equipage of ivory traders coming to the Angolan coast from Malebo Pool between 1869 and 1873. Charles Jeannest, *Quatre années au Congo* (Paris: Charpentier, 1883), ill. facing p. 142. Other examples were noted by von François along the middle Tshuapa in the 1880s and by the missionary Joseph Clark in the area around Lake Tumba in the 1890s. See Kurt von François, *die Erforschung des Tschuapa und Lulongo* (Leipzig: Brockhaus, 1888), p. 159; and FM Acc. # 33916, 33917, and accession information.

[70] Interviews, Georges Boleteni, 21 and 30 July 1989.

[71] Interview, Lotoy Lobanga, 27 July 1989. *Ebali* was a Bobangi term for the best or superlative person or thing. John Whitehead, *Grammar and Dictionary of the Bobangi Language* (London, 1899).

[72] Interview, Lotoy Lobanga, 26 July 1989.

[73] Interview, Elombola Boponde, 29 July 1989.

[74] Kriger, museum notes, technical data from the 107 examples I surveyed. Most had single midribs, some had double ones.

[75] Interviews, Nsabonzo Bilanga, 31 July 1989; Lotoy Lobanga, 27 July 1989; Elombola Boponde, 29 July 1989.

[76] The visual evidence is supported by lexical evidence, which suggests this sword type was generally and well established throughout Ngombe- and Lomongo-speaking areas.

[77] This photograph was taken in 1913 in the Territory of Inongo, on the eastern shores of Lake Mayi Ndombe. Van der Kerken, *l'Ethnie Mongo*, Vol. I, Plate XXVI, no. 116.

[78] MRAC, Department of Ethnology, EPH 5739 and 5742.

[79] MRAC, Department of Ethnology, EPH 1278.

[80] Van der Kerken, *l'Ethnie Mongo,* Vol. 1, Plate XI, No. 47; MRAC, Department of Ethnology, EPH 3022, EPH 2.293, EPH 2.294, and EPH 12.197.

[81] Hulstaert defines the term *ikákáláká* as a type of *ingóndá* knife with an especially narrow blade; and *byongi* as an *ingóndá* with numerous or long ornamental protrusions. Hulstaert, *Dictionnaire Lomongo-Français*, Vol. I, pp. 468, 779.

[82] Lumbwe Mudindaambi, *Objets et Techniques de la Vie Quotidienne Mbala*, 2 Vols. (Bandundu: CEEBA, 1976), Vol. I, pp. 9–11; Idem, *Dictionnaire Mbala-Français*, Vol. II, pp. 308–9, Vol. III, p. 663.

[83] Mpama and Lia chiefs and *nkúmú* title holders, for example, and other notables. MRAC, Department of Ethnology, EPH 7207, 13700, 3022, 5491, 14542, and 2.287; and Wauters, "l'Institution du 'nkum'," p. 157.

[84] MRAC, Department of Ethnology, EPH 5709, 5736, 2.2940.

[85] H. D. Brown, "The Nkumu of the Tumba," *Africa* 14 (Jan. 1943), pp. 436–37. Hulstaert asserts that only a man who, by birth or by influence, was recognized as *primus inter pares* of a group of "clans" could award the *bolúmbú* title to one of his wives. Such a titled woman was not the same as either the preferred wife or senior wife. Hulstaert, *Mariage*, pp. 348–54. See also Vansina, *Introduction*, pp. 85–86.

[86] Brown, "Nkumu"; Hulstaert, *Dictionnaire Lomongo-Français*, Vol. II, p. 1494. It is not clear at all to me, and other scholars have not questioned, whether the term "wife" was being used figurally or literally here.

[87] Interview, Lotoy Lobanga, 25 July 1989; Group interview, Lopanzo blacksmiths Nsatale Iseke, Ikete Boteni, and Bakonongo Bekoko, 27 July 1989; Group interview, elder women in Iyenge, Nsombo Bolaki, Bakombe Bokongo, and Inoko Bololo, 28 July 1989.

[88] See Daniel Vangroenweghe, *Bobongo: La grande fête des Ekonda* (Berlin: Dietrich Reimer, 1988).

[89] Personal communication, Lokongo Lombewo, 1 August 1989. See my photograph of a pigmented knife in Lopanzo in Kriger, *Ironworking*, p. 319. Examples of such pigmented knives exist in museum collections, such as MRAC Acc. # 59.48.422.

[90] See the photograph published in Célis, "Fondeurs et Forgerons Ekonda," p. 129.

[91] Hulstaert, *Dictionnaire Lomongo-Français*, Vol. I, p. 564.

[92] Von François, *die Erforschung,* fig. p. 72 and p. 159; Glave, *Six Years,* p. 151; Van der Kerken, *l'Ethnie Mongo,* Vol. I, Plate XX No. 89 and Plate XXI No. 94.

[93] Incised parallel lines on the surfaces of iron blades was a feature of products made in the northern and eastern Zaire basin, and appears to have been a technique transferred to blacksmiths along the middle Zaire from the Ubangi River basin southward.

[94] Coquilhat, *Sur le Haut-Congo,* p. 155; Johnston, *Grenfell*, p. 434; von François, *die Erforschung,* p. 159; Thonner, *Dans la Grande Forêt,* Plate 28; Masui, *Guide*, p. 101; Frederick Starr, *Congo Natives* (Chicago: Lakeside Press, 1912), Plate XLIX.

[95] Interview, Elombola Boponde, 29 July 1989; Lotoy Lobanga, et al., 27 July 1989.

[96] I identified only just over twenty examples in the collections of the Ethnology Department of the MRAC, and found no examples illustrated in nineteenth-century written sources. Of the examples I have surveyed, many had an ogee midrib, which was a form of midrib designed for airborne blades.

[97] Interview, Lotoy Lobanga, 26 July 1989; Torday and Joyce, *Notes Ethnographiques* (1922), p. 167; Hulstaert, *Dictionnaire Lomongo-Français*, Vol. I, p. 240.

[98] Kriger, field notes, July 1989.

[99] Kriger, field notes and photographs, 16 July 1989.

PART IV

IRONWORKERS IN WEST CENTRAL AFRICAN SOCIETY

8

The Blacksmith's Mystique Unveiled: Identity, Ideology, and the Social Prominence of Ironworkers

In central Africa, as in other areas of the world, ironworking offered career opportunities that could add new and significant facets to men's identities. Understanding the creation of social identities, and how complex and multilayered they can be for a single individual, is an important prelude to understanding broader historical trends and events. It has to do with how, at given times and places, people come to see themselves as aligned with this or that group of people, based on ethnicity, nationality, class, religion, and gender, and with what these groups then do, once they are formed. In addition to the more commonly acknowledged factors above, there is another factor in identity formation that is the subject of this study. It is the importance of one's work in creating commonalties and ties among individuals, with occupational social groups often forming according to whether the work is skilled or unskilled or somewhere in between. Here we have seen an artisanal occupation—ironworking—a set of skills, techniques, and products around and through which men formed a distinctive position for themselves in society, a facet of male identity that made them highly valued, well recognized, and admired throughout an ethnically diverse region.

Working iron was not simply an economic activity. As with all forms of work, it was human labor and much, much more—in this case presenting men with the possibility of inhabiting a very particular place in society. Just what that social place could be like, its size and shape, how inviting and hospitable it might be, changed over the centuries with changing historical conditions. It was contingent on what specifically the work was at a given time, how the work was carried out, where and by whom, and how it then came to be viewed by others as a result. Men who chose to become ironworkers in nineteenth-century west central Africa probably had a greater range of opportunities than any of their predecessors ever had. And like their predecessors, they used their occupation to create new ways of defining and presenting themselves as worthy and new ways of earning the admiration of others. The more successful ones generated economic wealth, social complexity, and above all the intangible social capital of prestige and respect.

Surveying ironworking regionwide in nineteenth-century west central Africa offers a view of the different sorts of workshops developed and strategies followed by male artisans during a time of relative prosperity. Moreover, in tracing some of the histories of changes made in their tools, techniques, and products, it becomes apparent that they had established informal networks among themselves and, just as importantly, reputations and contacts that reached into many other segments of their societies. As they did so, they used their occupational identity to rise above and beyond their local affiliations and form broader ones based on wealth and privilege. Master ironworkers, even though they were situated in scattered rural and urban workshops, formed a skilled, multiethnic elite of men who were sometimes regionally famous. Men in this position, rich and prominent, might then make alliances and hold social and economic interests in common with political leaders, powerful warriors, and other influential patriarchs.

Peoples all over the Zaire basin region registered their high esteem for ironworkers in popular proverbs, songs, and oral traditions. Surveying some of the specific, individual references is illuminating, for they recommend among other things staying on good terms with a blacksmith and being indulgent with his children. They characterize the smith as a man who is clever and talented, who expects to be compensated well for his work, and who, as the maker of tools and weapons, may be entitled to a share of the harvest and hunt.[1] They tend to refer quite precisely to blacksmiths, less so to smelters. Along similar lines, oral traditions and testimonies from the Kuba Kingdom and the town of Lopanzo place an emphasis on famous smiths or smelter-smiths. Of those central African ironworkers who did attain prominence in the nineteenth century, and of course not all of them did, the combined oral evidence indicates that most were blacksmiths, men who worked at the forge, the makers of semifinished and finished products. To better appreciate why this would have been so, we must return to have a final look at all of the work the occupation entailed, from workshop to consumer.

WORK, STATUS, AND CULTURE

Ironworkers' identities took on one shape or another slightly different one depending in large part on the types of social relations they were able to develop around their work. It mattered a great deal to a man's professional networks and contacts whether he specialized in either smelting or smithing or both. Division of the occupation into the specializations of smelting and smithing is an essential precondition for understanding workers' options and strategies, and thus what kinds of ties they were likely to create and with whom. Comparing these specializations systematically is a useful starting point for trying to better understand why it was that blacksmiths, especially, were remembered and honored over the past several centuries as historical figures. For smelters, their most complex relations had to do with primary production, especially the recruitment and management of labor, while smiths, who were small-scale entrepreneurs, could enter into highly complex social networks within the many commercial and prestige markets that existed for ironwares during the nineteenth century. Those who combined the two specializations were wise to do so.

Master ironworkers were masters of their own workplaces, for there is no evidence so far of any direct ownership of workshops by either merchant capital or political leaders. The decentralized, small scale of production was the result of ironworkers' control over the most important aspects of their work. They designed and made their own capital equipment and regulated access to the technical knowledge and skills necessary for using it. Impermanent furnaces and portable forges were particularly advantageous, for they offered geographical mobility to both smelters and smiths in setting up their workshops or relocating them. In general, ironworkers as a group enjoyed a great deal of autonomy in their working lives, and that brought their occupation a high labor status.

There were differences, however, in how smelters and smiths were able to supply their workshops, and in this respect it was blacksmiths who had some clear advantages. Given the specific conditions of trade in the region during the nineteenth century, smelters faced greater difficulties in gaining access to their raw materials. Not all smelters were able to have unencumbered and unlimited access to the large quantities of ore and charcoal their furnaces required. Only those who belonged to houses with land rights were so fortunate. Most had to negotiate with leaders who held rights to mines, arranging to pay them a fee, perhaps a share of the bloom. In contrast, blacksmiths were much more likely than smelters were to acquire their raw materials, especially refined iron, through trade. Numerous iron currencies were in circulation, overshadowing what trade there was in iron ore, the latter being a trade that developed only in specific places at certain times, such as the Lopori River basin in the late nineteenth century.

Even more significant differences were to be found in the area of labor requirements. The biggest bottleneck of primary iron production, especially for smelters who wanted to increase their output per season, had to do with the labor involved in preparing for a smelt. The washing, sorting, crushing, drying and/or roasting of ore was highly labor intensive, and recruiting and managing workers must have been especially challenging during the nineteenth century, when economic expansion was creating pressing demands for labor in other sectors of the economy. Examples of higher volume smelting, such as Ndulo in Angola, show that the key to these production levels was dependent and unfree labor, probably mostly women and children. Such examples of intensified smelting were not the norm, however, meaning that production levels overall remained generally low. In contrast, the labor needs of smithing workshops were far less burdensome. Bloomery iron was ready to work, and semiskilled tasks at the forge were carried out by male apprentices, a form of dependent labor quite different—and probably more consistently willing and manageable—than the family and slave labor used in smelting centers.

Finally, there were some important environmental constraints on smelting. Despite the efforts made by smelters to improve the output of their furnaces, the potential aggregate year-round volume of iron production was decreased by their adaptation to seasonal conditions. They restricted their work schedules mainly to the dry season periods, when hydrated ores could be dried or roasted and charcoal fuel could be kept relatively free of moisture. Smithing, however, was not subject to the same restrictions, and there were blacksmiths who were known to work year round.

Based on nineteenth-century production conditions, smithing appears to have been the more attractive, less risky, and most easily entered of the two specializations, which goes a long way toward explaining why it was more widely practiced. As primary manufacturers, smelters faced greater difficulties in consistently supplying their workshops with raw materials and labor. As secondary manufacturers, the makers of value-added goods, smiths were in a better position to amass greater wealth by generating higher profit margins relative to smelters. Moreover, they did not owe shares of the final products of their workshop to their assistants or raw materials suppliers, as smelters might.

Even more important advantages associated with smithing had to do with the particular distribution and consumption patterns that existed during the nineteenth century. A great boon to smiths were the cultural values of the societies they lived in, which assured them multiple markets for their goods and a variety of options for experimentation, innovation, and advancement. Iron was considered a semiprecious metal, and we have seen over and over again how and where it operated in many forms as currencies and stores of wealth. Hence a smithy, if well stocked with supplies of iron for future use, could serve also on occasion as a bank, offering loan and currency conversion services to the community.[2] Blacksmiths also equipped and outfitted the wealthier segments of society. The many different luxury products they invented and replicated at the forge served the tastes and fashions of powerful and prominent men, consumers who had the greatest purchasing power. All in all, the opportunity cost for smelting was high relative to smithing,[3] unless smelters diversified their workshops by adding a forge to produce semifinished iron goods.

Although the material base of ironworking—primary production—is one important factor underlying the special status of ironworkers, it alone does not tell the entire story, and does not explain the full meaning of the work and the ways it flourished, especially over the nineteenth century. Smithing, particularly finish work, is a distinctly different working process, involving other important factors in addition to equipment, raw materials, and labor relations. In the oral societies of precolonial central Africa, certain cultural values and their visible realization and expression in the form of metal objects were also fundamental keys to the high status of blacksmiths. In other words, technology, production, and culture each played an important part in the making of ironworkers' privileged social position.

Central African ironworkers, blacksmiths particularly, were admired and respected for many reasons, some now apparent, some probably still elusive. Surely one major reason, though, was their skilled handling of and control over such a valuable material, iron, a metal that was so necessary, desirable, and beautiful to so many important people. In locales all over the Zaire basin region, iron was carefully conserved and preserved, even the smallest pieces of it protected or worn as jewelry. It could circulate in currency markets, be turned into a commodity, then used, with the remaining portion eventually welded together with other bits of iron into another commodity or currency unit. Smiths were manufacturers, and also the repairers and recyclers of precious iron goods. Many blacksmiths were revered as the makers and changers of money. Others became highly skilled masters who grew wealthy and well known, making, among other things, coveted insignia, prestige goods, and works of art for the leading men in their societies. Smelters and their low production levels created the scarcity and high worth of iron metal, a regard that could be seen permeating cultural values all over central Africa. These values, in turn, spawned the markets for currencies and luxury goods in iron. Iron was valuable indeed, and it was smiths above all who were able to capitalize on this.

IRONWORKERS AS A SOCIAL GROUP

Like the households and villages of central Africa, ironworking workshops were multiethnic products of history that were cloaked in the guise of a family and lineage ideology.[4] Oral testimonies about the training of young ironworkers were often generalized, and when taken literally, created the impression that workshops were local and lineage-based. There, it seemed, the smelter or blacksmith relied on his own and family labor to make goods solely for his neighbors' subsistence needs or for bridewealth payments to lineages nearby. In normative accounts of how work was carried out and what products were made, it was the customary, local dimensions that were stressed. Training relationships, too, were described in such a way that the occupation seemed to have been kept strictly within the family and lineage. Ironworkers' training supposedly began and ended in their youth, the requisite skills, along with basic equipment, passing from father to son or uncle to nephew.

Ironworking could indeed be inherited, that is, entry into the occupation was often a function of filial or avuncular relationships between males. But that does not present the complete picture of entry into the trade. Appended to these normative accounts could be all kinds of exceptions based on lived experience, where it was revealed that ironworkers would accept for training men and boys who were not their own relatives. Sometimes these exceptions were rationalized by describing the trainees as having shown particular promise or talent. But it is not unreasonable to infer that blacksmiths, especially, might have arranged to train just about any boy if the price was right. In some cases, wealthy owners of slaves made deals with blacksmiths to have certain of their male slaves trained to work iron, and there were examples also of Twa clients learning to smith. Moreover, there were times and places where essential ironworking tools such as hammers and anvils circulated through bridewealth networks and in currency and commodity markets, opening up access to equipment, making practice available in ways other than through family inheritance. The prominence achieved by ironworkers was theirs not by birthright, but by the talent, persistence, and acumen individual men exercised in manufacture, management, and marketing.

What is most crucial to understand is that ironworking was a lifelong occupation. Smelters, and especially blacksmiths, entered the occupation in their youth, but they also then trained and retrained throughout their lifetimes. Some ironworkers probably pursued the craft in much the same way as their masters had. Others were more ambitious, and it was they who were largely responsible for initiating professional ties between ironworkers in different locales and language groups. Innovation and retraining drove change in ironworking that could pass from one place to another through the informal networks that ironworkers created, especially around and between masters. Smelters sought new techniques and furnace designs from other masters, and blacksmiths joined smelting teams to learn about making furnaces and operating them. The appearance of new, more desirable products on the market prompted smiths to try to copy them or to travel to workshops far away to acquire the necessary skills and techniques for making them. Well-known ironworkers also were sometimes encouraged to immigrate, and when they did, they brought their knowledge and expertise with them, henceforth passing them on to trainees in their new communities.

Normative descriptions of how ironworkers were trained were incomplete, presenting only one, very limited perspective, one that stressed continuity of the occupation within a local, familial context. Patterns of change in ironworking across family and ethnic lines were revealed through my analyses of other evidence, mainly the metal products made in central African workshops and lexical terms related to them. Many initiatives taken by ironworkers to upgrade their skills and retool their workshops were registered as discernible changes in the products they made and the languages they spoke. This evidence also clarified the limitations of oral testimonies and traditions. Some changes in tools, techniques, and equipment may have occurred incrementally and slowly, and so had not been consciously experienced and consigned to memory; others were only recalled generically, as a chain of interethnic relationships. Recovering this new evidence and presenting it here shows how the history of ironworking in central Africa was shaped not by the existing social structures of houses and families, but by the career decisions of individual blacksmiths themselves.

Ironworking was an occupation open to men and men only, although women were sometimes recruited to work in mining and preparing the ore for a smelt. Strict prohibitions forbidding women to come within view of a working furnace or to handle the blacksmith's hammers and anvils succeeded in keeping the most crucial technical knowledge and skills within the circles of select men. It was not just that women themselves should not have been allowed entry into such a potentially lucrative occupation. Secrecy served as a patenting system, and women, who married out and usually resided in their husbands' villages or towns, were possible breaches in the system. In other words, their loyalties were divided—between their own and their husbands' families—and they could pass along valuable secrets to other men. This principle was recognized and highlighted in oral traditions, for example, in a Bushoong cliché about the invention of kingship. In variants of the episode describing a competition between rivals on the plain at Iyool, a woman betrays a metalworking secret from her own family to that of her husband's, or vice versa. The point is that exogamy and virilocal residence rendered their own sisters, daughters, and wives unreliable in men's eyes.

Ironworkers claimed to be "insiders" while creating a complex occupational identity, an identity that was accorded high status and respect. Some of them transcended their important local and ethnic affiliations, making their work a vehicle for cutting across groups formed by kinship or other dependent relations. They consistently defined their work by gender, less so by ethnicity. However, they were not organized formally or cohesively into guilds or quasi-brotherhoods, as far as is known. Nor were they formed into endogamous "castes," as some ironworkers were in West Africa, for they could and did marry into elite and even royal families. But neither were they isolated individuals working independently of one another in small-scale, subsistence-oriented workshops. Their ambition led them to form working ties and professional relationships amongst each other to further their training and careers. They were also recognized in their societies and singled out as a group of experts who through their manual and social skills could acquire power, the profane power of wealth and privilege. In the political machinations and maneuverings leading to the creations of polities and states in west central Africa, it is no longer puzzling to find that individual ironworkers were among the men involved.

IRONWORKERS AS CULTURE HEROS

At crucial, pivotal moments during some of the working processes, especially in smelting, ironworking was cloaked in a garb of ritual and cosmology. These were moments of extreme risk, when smelters and smiths were exposed to the limits of their control, be it over the ultimate outcome of a smelt, for example, or the challenge of welding together relatively large masses of iron to form a hammer or anvil. Rituals and divination procedures that were carried out at smelts and in smithies served a number of purposes, some of them not unlike the religious consecrations of factories and spontaneous prayers for success that have been performed in workshops elsewhere in the world. Moreover, they integrated the work and workers into local ideologies, by underscoring the potency of current religious beliefs.

Ritual also served other very practical purposes. Medicines and ritualized behaviors deployed at times during ironworking processes, especially in smelting, often were designed to create physical and metaphysical conditions that would make failure less likely. They also could be used afterward to minimize the impact of failure. For when failures did occur, explanations based on negative spiritual forces or improper ritual behaviors deflected blame and criticism away from the expert in charge. Thus while ironworkers' names came to be recognized and enhanced by the very real successes they did achieve, rituals came to their aid in a number of ways, including protecting their reputations.

Religious embellishments of the working processes indicate just how important the successful completion of the work was; however, they say very little else about the work itself. Certainly when ironworkers were closely associated with religious specialists or when they themselves became diviners, that did add to their own respectability and power. But however deep their religious involvement was, that alone cannot and does not account for their prominence in history and collective memory. Rituals associated with ironworking have long been somewhat of a distraction to scholars, drawing attention away from the ongoing daily working processes, especially those of blacksmiths. Rather than trying to see ironworking through its symbolic roles in cosmology and ritual, knowing the history of ironworking more fully can help in understanding the reasons behind why such symbols were created and why they were attached to master smiths in so many different places.

Blacksmiths were among the esteemed historical figures remembered in west central Africa, but not because they were magician-priests, steeped in powers of the supernatural. Already it is abundantly clear that ironworkers were a prominent group of men who could and did become members of secular political elites. They also could justly be viewed as potential culture heros, and the case studies presented in this book show how and why. For good reason, they were known as innovators within their societies, men who invented new products or works of art, and who could learn to replicate imported goods from elsewhere. Their networks were the conduits for information and technology transfer, whereby ideas, tools, and techniques could be absorbed in new places. Smelters and blacksmiths moved and traveled, trained and retrained, bringing about changes in central African languages and societies along the way.

Reputations of the most cosmopolitan, clever, successful, and wealthy ironworkers could last long after their deaths, in part through some of the investments they made. Smelters and smiths did not simply rise to prominence on their own achievements in their workshops. They were supported by a productive base consisting of the labor of women, Twa clients, and slaves. The wealth of smiths especially, like the wealth of big men and traders, was invested in people, which allowed them to create large followings, households, and families. Sometimes it allowed them to move and become the founders of entirely new communities, where they then would have the authority to administer land rights. Investments in people were investments in their futures as historical figures, for it was through their descendants that they could be traced and remembered in local oral histories.

The passing of generations turned prominent men into heros and heros into symbolic, mythological figures. For while ritual embellished the manual and managerial skills of ironworkers as they worked, it was time and its duration that compressed, distilled, and amplified the memories of their reputations. The legacies of blacksmiths were not only their progeny but also their products, especially the most important ones, their own tools and other masterworks. Iron was precious, long-lasting if well cared for, and worthy of preservation in and of itself. Treasuries held by families, houses, chiefdoms, and kingdoms contained all sorts of metal goods kept for all sorts of reasons. Ironworking tools and former currencies were stores of wealth as well as the basic start-up equipment for a future smithing workshop. Certain kinds of products also were the demonstration models and prototypes used by blacksmiths to teach apprentices or to remind themselves how to make them. In some cases, highly valued prestige objects and works of art were kept as treasures, and they were displayed proudly by descendants on ceremonial occasions—as visual evidence of past mastery.

Just as the ironworkers themselves and their workshops were living accretions of knowledge and skills that came from many times and places, treasuries were the accumulations of more than one blacksmith. The works ascribed to famous blacksmiths of the past were not necessarily all made by those very same individuals. Even some of the works reputedly made by the master Bana in Lopanzo may have been older treasures or items of inventory that had been made by his own masters or predecessors. Certainly not all the great masterpieces said to have been made by the Kuba smith Myeel were the work of a single pair of hands. There are limits to memory, and so in the oral traditions, single names were selected to stand for the collective achievements and legacies of the most successful ironworkers. In leaving behind the mechanisms by which they could be remembered and turned into myths—progeny and masterworks—blacksmiths assured for themselves a place in central African history and a mystique of their own making, recalled today as the "pride of men."

NOTES

[1] For one particularly rich sampling, see Korse, "La Forge."

[2] One specific example mentioned briefly in the ethnographic literature is the case of smiths in Kole (southern Mongo), who loaned out local iron money and charged fees of up to 20 percent. L. Liegeois, "Notice sur le régime social des Basongo Meno de Kole," *Bulletin des Juridictions Indigènes et du Droit Coutumier Congolais* 9, 1 (1941).

[3] Opportunity cost is the income one loses by doing one kind of work rather than another. Harms has stated that in general, for the middle Zaire basin, the opportunity cost for craft production was low, because it fit in well with the agricultural cycle. Harms, *River of Wealth, River of Sorrow*, p. 60. This observation is useful as a general statement, but it surely will be revised as more detailed historical studies of craft production are carried out. It certainly does not apply very well to nineteenth-century blacksmiths.

[4] My work supports the findings of Vansina, that "lineage" and "clan" were ideological constructs, and that the social reality was much more complex. See Jan Vansina, "Lignage, Ideologie et Histoire en Afrique Equatoriale" *Enquêtes et Documents d'Histoire Africaine* 4 (1980); and idem, *Paths*.

Epilogue: Colonial Rule and the Decline of a Craft Tradition

These stories of central African ironworkers reveal them, in their multifarious communities and workshops, as a special and revered social group. It should seem now strange that today much of the diversity and dynamism of their occupation is no longer evident. Hence the stories of ironworkers' decline are also important, though yet to be told in full. Like their homologues in Europe and America, this once robust artisanal group saw a steady diminution in their own numbers, their incomes, and their status over the course of the twentieth century. But the similarities only go so far. British artisans, for example, though eventually doomed by transformations in their highly industrialized economy between the two World Wars, had managed to play significant roles in both industrial development itself and in the creation of organized trade unions.[1] In Africa, the decline of artisanal groups occurred in very different economic conditions and within the peculiar context of colonial rule, which made it very unlikely that ironworkers would be able to leave behind a comparable historical legacy. Artisans in the Zaire River basin encountered industrial capital much more abruptly and recently, and on top of that had to contend with alien, racist governments and a host of policies and initiatives intended to rapidly restructure their societies and economies.

One result that has already been noted in the literature was the ending of bloomery iron smelting throughout most of the continent by the 1950s. Just why smelters stopped reproducing themselves, however, is not entirely clear. An oft-repeated explanation is that indigenous African manufacturing, especially during colonial rule, was rendered obsolete and displaced by more competitive, cheaper goods imported from overseas. And indeed, one can find anecdotal evidence that supports this general claim. But it is an insufficient explanation for the decline of all aspects of smelting and smithing throughout this region in central Africa, an area as large as western Europe. Such a logical if highly reductive explanation might be credible if local iron manufacturing had been organized around producing just a few common, utilitarian products in volume quantities, products that could be copied in

European factories and easily replaced. This particular model of production organization was, however, relatively unusual.

Those few examples where one single product was manufactured in large quantities were the smelting and refining workshops that turned out hoe blades, such as those around Ndulo, or Fipa, which lies outside our study area. And even then, the story of when, how, and why the smelting furnaces stopped operating was not a simple one with a single cause. One of the best attempts so far to tell such a story focused specifically on the *iluungu* shaft furnaces and large hoe blades made by Fipa and Mambwe ironworkers in southwest Tanzania. There, the locally made hoes were indeed replaced eventually by cheap imports, but only after a British manufacturer finally succeeded in replicating a hoe according to the specifications of the preferred local design. Moreover, the process of decline occurred over a forty-year period, and competition from overseas products was not the only factor involved. Other factors included colonial policies having to do with forest conservation, which led to the prohibition of charcoal making after World War II. Equally important were policies affecting labor; as men sought other work opportunities in the formal colonial economy, smelters experienced increasing difficulties in labor recruitment. High levels of male labor migrancy also prompted the introduction of ox-drawn ploughs, which decreased men's labor, increased women's labor in agricultural production, and reduced the demand for hoes. Blacksmiths thus made fewer products and felt obliged to charge relatively low prices, while at the same time they were forced to depend more and more on scarce and costly scrap iron to supply their workshops.[2] In short, there were interrelated changes in agricultural technology, in labor allocation and opportunities, and in the availability of fuel and iron supplies that, when added to the importation of British-made hoes, brought about an end to smelting and a decline in smithing in this particular case.

Most smelting and smithing workshops in the Zaire basin region, though, were more diversified than the ones in Fipa, and were not only producing farm implements but were oriented toward the variety of markets that I have identified and described in this study. For that reason, I am prompted to offer a few more general comments here as both an ending for this book and as a beginning, looking toward future research projects aimed at understanding the transformations of precolonial patterns of work during the era of colonial rule. Given my conclusion that the interrelated and interwoven histories of ironworking demonstrate that workers enjoyed autonomy and high status, and that their production was diversified, the occupation's decline seems all the more striking. It can be viewed as a measure of the wide-ranging impacts of colonial policies and attitudes on central African society at all levels. A craft tradition that had been maintained over two millennia, that had survived, even flourished, during a violent and turbulent nineteenth century, came to an end only very recently, between 1920 and 1950, as European rule became consolidated into deeply paternalistic colonial regimes.

The formal period of colonialism in tropical Africa, namely the 1920s to 1960s, has been aptly characterized as a time when the range of choices and options available to individual African men and women diminished. Ironworking was an occupation built out of diverse strategies, options, adaptations, and innovations, one which afforded great worker autonomy and mobility. By examining the decline of ironworking, we can come to understand more concretely just what this loss of choice could mean to many men, and can highlight the variety of policies and attitudes that lay behind it. Certainly, the factors mentioned above with respect to ironworkers in southwestern Tanzania had their parallels in the Belgian Congo, the major colonial regime in the Zaire basin region. Other factors not usually considered relevant to ironworking *per se*, include the introduction of new definitions of and values for work, different avenues for pursuing male success and prominence, centrally produced and managed currencies and the devaluation of iron, the *pax Belgica* and its centralized state control over armaments and weapons, and new symbols of status and wealth. These factors were especially relevant to blacksmiths who, unlike smelters, continued to reproduce themselves, though in smaller numbers and for work that was almost entirely changed from what it had been before.

Work in general was fundamentally transformed with the imposition of a free wage labor ideology along with forced labor practices during colonial rule.[3] Among what was left behind, what these forms of labor supplanted, were untold numbers of skills, tools, and work strategies that formerly had been well integrated into central African communities and their cultures. Artisanal work brought, in addition to income, an aspect of identity, and offered to the artisan a valued position in society. Hence, the undermining of this work is an important part of the colonial story. During the twentieth century, it was not only freed slaves, or displaced farmers, peasants, and herders who were forced to turn to wage labor or work on plantations, but independent artisans as well. In other words, the full range of precolonial forms of work and income generation must be taken into account and understood first before modernizing processes such as proletarianization can be seen more completely in concrete and human terms.

Ironworkers had been powerful members of their societies in precolonial times, patriarchs who could and did exploit women, children, and other men in their pursuit of success. But they, too, experienced the violence of empire and were themselves exploited as their occupation atrophied over the course of just a generation or two. Some blacksmiths still continue working today, though mostly in rural areas and in the informal economy, making some few tools and weapons for the hunt, and engaging in repair work. The social authority and wide scope of maneuverability enjoyed by ironworkers of the past exist no more and now smiths contribute primarily to subsistence production. And so it is the observations made of them and their work in the twentieth century that must be responsible for a commonly held notion that smiths made metal goods only for local consumption. That description might be accurate for many workshops in recent times but it certainly does not hold for the ones that were in operation one hundred years ago.

A major blow to blacksmiths was the loss of several important regional markets for which they had worked. One of these was the currency market. Currencies imposed during the earlier Congo Free State regime did have some dramatic effects on metalworkers, but not nearly so completely as during the Belgian period. After 1885, Leopold's administration employed the *mitako* unit of imported brass wire in payments of fines and wages, which affected different metalworkers in different ways. It must have dealt a blow to the copper smelters in the lower Zaire area and in the copperbelt, but it was in some ways a boon to smiths who were still able to convert currencies and cast or rework them into stores of wealth. However, with more rigorous efforts after the turn of the century to impose a modern cash economy based on the franc,[4] with coins and paper money issued in Belgium, ironworkers' involvement in making and converting units of currency diminished. By the time of the Second World War, many transactions for purchase of goods, paying fines, and payment of bridewealth were being made in the colonial cash currency,[5] and both the value of iron as a metal and its circulation velocity had gone down. This important aspect of blacksmithing work and a source of its prestige eventually disappeared completely.

Certain kinds of weapons continued to be made, though in far fewer numbers. It was evident in Lopanzo, and probably so in much of rural Zaire, that blacksmiths still maintained the skills for producing various types of spear blades, arrow blades, and harpoons for hunting game. To some degree, other blade forms that had been designed specifically for use in human combat were rendered obsolete by modern advances in weapons technology, though machetes and poisoned arrows can still be very effectively deployed. Some weapons designed for combat are made and used today for ceremonial display, though in certain cases their forms have been modified and are no longer structurally sound enough for that original purpose.

Another market that was affected was the one for metal luxury goods modeled after various insignia of office, products that had prompted so much innovation and retraining by blacksmiths in the nineteenth century. Medals were used from 1889 onward by Congo Free State officials to reward local leaders for service and loyalty to the regime,[6] and these were turned into new insignia of office as the Belgian colonial administration instituted positions for approved chiefs in 1910. Many chiefs selected by colonial officials to hold such positions might have been ineligible, or at best, not very strong or well-supported candidates for local leadership according to customary law.[7]

Nevertheless, they were designated as legitimate government officials, their legitimacy visible to all and verified by the wearing of fancy government medals that were made in Belgium. Hence not only were the historically valid procedures for selecting leaders undermined, so too was the political potency of locally-made insignia. Although some blacksmiths today retain the skills necessary for making some of these older prestige objects, they usually only make such products for sale to foreign tourists. The rationale and impetus for innovation, however, have gone.

Meanwhile, memories of the heyday of ironworking persist, as do the treasuries of metalworking tools and products, serving to acknowledge what past masters did, and generating a particularly potent nostalgia for the times in which they lived. The very real privilege that ironworkers once enjoyed has thus been transformed into a collectively held pride in the precolonial past. As such, the craft tradition these men made, like artisanal traditions elsewhere in the world, has become a symbol of cultural continuity and independence, a means of maintaining dignity under difficult conditions, and a source of hope for a future renewal.

Younger generations of men have sought out new avenues of opportunity during this century, not easily finding success. One of the luckier ones is Samba wa Mbimba N'Zinga Nurimasi Mdombasi, who in 1972, at the age of sixteen, had to drop out of school for financial reasons and struck out on his own. He moved to Kinshasa where he became an apprentice sign painter, established his own workshop three years later, and then turned it over to his assistants so that he could devote full time to painting. In 1979 he adopted the professional name of Chérie Samba, and took part in his first group art exhibition in Europe. Now an internationally known artist who has had solo shows of his work in Paris, New York, and Chicago, his paintings employ image and text to engage the viewer in current issues and social commentary.[8] It is not incongruous at all that such a talented, ambitious, and cosmopolitan man from central Africa describes himself proudly as the son of a blacksmith.

NOTES

[1] Hobsbawm, "Artisans and Labour Aristocrats?" in *Workers: Worlds of Labor.*

[2] Marcia Wright, "Iron and Regional History: Report on a Research Project in Southwestern Tanzania," *African Economic History* 14 (1985), pp. 155–58.

[3] Industrialized mining became the mainstay of the Belgian Congo's economy, and an early focal point in the development of wage labor. For a look at how the work and workers were organized, and the impacts on workers' lives, see Jean-Luc Vellut, "Mining in the Belgian Congo" in Martin and Birmingham, *History of Central Africa*, Vol. 2. At the same time, colonial agricultural policies employed systems of forced labor. For these, and their effects on peoples in the Tshuapa River basin, see Samuel H. Nelson, *Colonialism in the Congo Basin, 1880–1940* (Athens, Ohio: Ohio University Press, 1994);

and for forced cotton cultivation, see Osumaka Likaka, *Rural Society and Cotton in Colonial Zaire* (Madison: University of Wisconsin Press, 1997).

[4] The legal money of the Congo Free State was, for Europeans, the franc from 1887 onward. *Bulletin Officiel de l'Etat Indépendant du Congo* No. 8 (1887), p. 118. For the rest of the population, however, taxes were to be paid in forced labor and produce, while they used the *mitako* in the cash sector.

[5] See Bogumil Jewsiewicki, "Rural Society and the Belgian Colonial Economy" in Birmingham and Martin, *History of Central Africa*, Vol. 2.

[6] *Bulletin Officiel de l'Etat Indépendant du Congo* 5 (1889), p. 134.

[7] Bogumil Jewsiewicki, "Belgian Africa" in *The Cambridge History of Africa* (Cambridge: Cambridge University Press, 1986), p. 467.

[8] See Miriam Rosen, "Chéri Samba, Griot of Kinshasa and Paris," *Artforum* (March, 1990); and Bernard Marcadé, "Chérie Samba," *Galeries Magazine* (February/March 1991).

WORKS CITED

PRIMARY SOURCES

Collections of Metalwork

Centre Aequatoria, Bamanya, Equateur, Zaire
Field Museum of Natural History, Chicago, Illinois, USA
Hampton University Museum, Hampton, Virginia, USA
Musée Royal de l'Afrique Centrale, Tervuren, Belgium
Museum of Mankind, London, England
National Museum of Natural History, Washington, D.C., USA
Private collections in Lopanzo, Equateur, Zaire
Redpath Museum, Montréal, Québec, Canada
Royal Ontario Museum, Toronto, Ontario, Canada

Word Lists and Vocabularies for Metalworking Terms from ca. 150 (Mostly Western Bantu) Languages.

These sources are in the possession of the author.

Interviews and Workshop Observations

North America

Douglas Hutchens, Engineer, Gerber Legendary Blades, Portland, Oregon, 21 December 1988.

David Norrie and Daniel Kerem, Blacksmiths, Stouffville, Ontario, 6 March 1989.

J.E. Rehder, Iron Metallurgist and Foundryman, Department of Metallurgy and Materials Science, University of Toronto, 13 March 1989, 27 September 1989, 28 March 1990, and 28 December 1990.

Group Interview: Ted Bieler, Sculptor; Daniel Kerem and David Norrie, Blacksmiths; J.E. Rehder, Iron Metallurgist; Z. Volavka, Historian; with the Hope Morgan central African metalwork collection, Department of Ethnology, Royal Ontario Museum, Toronto, 10 April 1989.

Assisted J.E. Rehder in laboratory iron smelts, *1988 Symposium on Archaeometry*, University of Toronto, 16-20 May 1988.

Central Africa

Georges Boleteni (b. ca. 1950), Lopanzo, Zaire, 30 July 1989.

Elombola Boponde, lineage spokesman (b. 1938), Lopanzo, Zaire, 29 July 1989.

Lotoy Lobanga, Master Blacksmith (b. 1924), Lopanzo, Zaire, 25 July 1989, 27 July 1989, 29 July 1989, and 1 August 1989.

Ndongo Bokoti, diviner-smelter (b. 1925), Lopanzo, Zaire, 22 July 1989.

Nsabonzo Bilanga, chef de localité (b. 1938), Lopanzo, Zaire, 31 July 1989.
Operations at smithy of Lotoy Lobanga, Lopanzo, Zaire, 25 July–1 August 1989. (Smiths: Lotoy Lobanga (b. 1924); Nsatale Iseke (b. ca. 1940); and Bakonongo Bekoko (b. ca. 1945).
Reconstruction of iron smelt, Lopanzo, Zaire, 15 July–3 August 1989.
Group Interview, lineage elders and spokesmen: Elombola Boponde (b. 1938) and Buka Bunkose (b. ca. 1910?); Lotoy Lobanga (b. 1924) and Bonyele Mbenga (b. ca. 1910?); Bopomwa Ninga (b. 1920) and Iyanza Liongo (b. 1910?); Lopanzo, Zaire, 24 July 1989.
Group Interview, elder women (all b. ca. 1910): Bangolo; Bakoliloka; Bakonabangala; Epandapanda; Eyalayala; Lopanzo, Zaire, 21 July 1989.
Group Interview, elder women (all b. ca. 1910): Nsombo Bolaki; Bakombe Bokongo; Inoko Bololo; Iyenge, Zaire, 28 July 1989.
All interviews except group interviews conducted by the author; group interviews conducted mainly by Dr. Eugenia Herbert.

Photographs

Joseph Maes (1913-14), Archives Africaines, Brussels, Belgium.
Emil Torday (1906-8), Royal Anthropological Institute, London, England.
Lang Expedition (1909-15), American Museum of Natural History (AMNH), New York City, USA.
Département d'Ethnologie, MRAC, Tervuren, Belgium.
Photographs by the author, Bamanya, Zaire and Lopanzo, Zaire, July and August 1989.

Unpublished Archival Sources and Field Notes

Archives, Section Historique, MRAC, Tervuren, Belgium.
Colonial Reports, Département d'Ethnologie, MRAC, Tervuren, Belgium.
Correspondence, Torday Congo Expedition 1906-8, Royal Anthropological Institute, London, England.
Field Notes, Torday Congo Expedition 1906-8, Royal Anthropological Institute, London, England.
Papiers Maes, Archives Africaines, Brussels, Belgium.
Research Reports, Institut des Musées Nationaux du Zaire (IMNZ), Kinshasa, Zaire.
Vansina Field Notes for Kuba Kingdom (1953-4) and (1954-6).
Walter Currie Papers, United Church of Canada Archives, University of Toronto, Canada.
William Sheppard Files, Hampton University Archives, Hampton, Virginia, USA.

Published Archival Sources and Contemporary Periodicals

Biographie Coloniale Belge
Bulletin Officiel de l'Etat Indépendant du Congo
Le Congo. Moniteur Colonial
Le Congo Illustré
The Missionary (1904)
Regions Beyond (1881-94)

BIBLIOGRAPHY

Ackerman, Kyle, David Killick, Eugenia Herbert, and Colleen Kriger. "A Study of Iron Smelting in Lopanzo, Equateur Province, Zaire." *Journal of Archaeological Science*. Forthcoming, 1998.

Agthe, Johanna and Karin Strauss. *Waffen aus Zentral-Afrika.* Frankfurt: Museum für Völkerkunde, 1985.

Atkins, Keletso. *The Moon is Dead! Give us our Money!* Portsmouth, NH: Heinemann, 1993.

Austen, Ralph and Daniel Headrick. "The Role of Technology in the African Past." *African Studies Review* 26, 3/4 (September/December 1983): 163-184.

Avery, Donald and Peter Schmidt. "The Use of Preheated Air in Ancient and Recent African Iron Smelting Furnaces: A Reply to Rehder." *Journal of Field Archaeology* 13 (1986): 354-357.

Baeyens, M. "Les Lesa." *La Revue Congolaise* 4 (1913-14): 129-205.

Barbosa, Adriano. *Dicionário Cokwe-Português.* Coimbra: Instituto de Antropologia, 1989.

Bastin, Marie-Louise. "Le Haut-Fourneau 'lutengo': Opération de la Fonte du Fer et Rituel chez les Tshokwe du Nord de la Lunda (Angola)." In *In Memoriam António Jorge Diaz.* Lisbon: Instituto de alta cultura junta de investgações cientificas do ultramar, 1974.

———. "Tshibinda Ilunga: A Propos d'une statuette de chasseur ramenée par Otto Schütt en 1880." *Baessler-Archiv*, Neue folge 13 (1965): 501-537.

———. *Statuettes Tshokwe du Héros Civilisateur 'Tshibinda Ilunga'*. Arnouville: Arts d'Afrique Noire, 1978.

———. *Art Decoratif Tshokwe*. 2 Vols. Lisbon: Museu do Dundo, 1961.

Bastin, Y. *Bibliographie Bantoue Selective.* Tervuren: MRAC, 1975.

Baumann, Hermann. "Zur Morphologie des afrikanischen Ackergerätes." *Koloniale Völkerkunde. Wiener Beiträge zur Kulturgeschichte und Linguistik* 6, 1 (1944): 192-322.

———. *Lunda.* Berlin: Würfel, 1935.

Bentley, W. Holman. *Pioneering on the Congo.* 2 vols. 1900. Reprint, New York: Johnson Reprint Co., 1970.

Berry, Sara. *No Condition is Permanent.* Madison: University of Wisconsin Press, 1993.

Birmingham, David and Phyllis Martin, eds. *History of Central Africa.* 2 vols. London: Longman, 1983.

Bisson, Michael. "Copper Currency in Central Africa: The Archaeological Evidence." *World Archaeology* 6, 3 (1975): 276-292.

———. "The Prehistoric Coppermines of Zambia." Ph.D. Dissertation, U.C. Santa Barbara, 1976.

Blackmun, Barbara and J. Hautelet. *Blades of Beauty and Death.* San Diego, California: Mesa College Art Gallery, 1990.

Bledsoe, Caroline and Gilles Pison, eds. *Nuptiality in sub-Saharan Africa.* Oxford: Clarendon Press, 1994.

Bohannan, Paul and George Dalton, eds. *Markets in Africa.* Evanston: Northwestern University Press, 1962.

Bolesse, F. "Essai Historique sur les Lusankani." *Aequatoria* 23 (1960): 100-111.

Borgeroff. "Les Industries des WaNande." *La Revue Congolaise* 3 (1912-3): 278-284.

Bourgeois, A. *Art of the Yaka and Suku.* Meudon, France: Chaffin, 1984.
Brelsford, W.V. "Rituals and Medicines of Chishinga Ironworkers." *Man* 49 (1949): 27-29.
Brooks, George. *Landlords and Strangers. Ecology, Society, and Trade in West Africa, 1000-1630.* Boulder, Colorado: Westview Press, 1993.
Brown, H.D. "The Nkumu of the Tumba." *Africa* 14 (Jan. 1943): 431-447.
Burrows, Guy. *The Curse of Central Africa.* London: Everette, 1903.
Cahen, Lucien. *Géologie du Congo Belge.* Liège: Vaillant-Carmanne, 1954.
———. *Esquisse tectonique du Congo Belge et du Ruanda-Urundi.* Liège: Ministère des colonies, Commission de géologie, 1952.
——— and J. Lepersonne. *Carte Géologique du Congo Belge et du Ruanda-Urundi.* Liège: Institut géographique militaire, 1951.
Cameron, Verney. *Across Africa.* 2 vols. London: Daldy, Isbister, & Co., 1877.
Célis, Georges. "Fondeurs et Forgerons Ekonda (Equateur, Zaire)." *Anthropos* 82 (1987): 109-134.
——— and Emmanuel Nzikobanyanka. *La Métallurgie Traditionelle au Burundi.* Tervuren: Musée Royale de l'Afrique Centrale, 1976.
Ceulemans, P. *La Question Arabe et le Congo 1883-1892.* Brussels: ARSOM, 1959.
Chaplin, J.H. "Notes on Traditional Smelting in Northern Rhodesia." *South African Archaeological Bulletin* 16 (1961): 53-60.
Childs, S. Terry. "Transformations: Iron and Copper Production in Central Africa." *MASCA Research Papers in Science and Archaeology* 8, 1 (1991): 33-46
———. "Iron as Utility or Expression: Reforging Function in Africa." *MASCA Research Papers in Science and Archaeology* 8, 2 (1991): 57-67.
———. "Style, Technology, and Iron Smelting Furnaces in Bantu-Speaking Africa." *Journal of Anthropological Archaeology* 10 (1991): 332-359.
——— and David Killick. "Indigenous African Metallurgy: Nature and Culture." *Annual Review of Anthropology* 22 (1993): 317-337.
———, William Dewey, Muya wa Bitanko Kamwanga, and Pierre de Maret. "Iron and Stone Age Research in Shaba Province, Zaire: An Interdisciplinary and International Effort." *Nyame Akuma* 32 (Dec. 1989): 54-59.
Claerhout, Adriaan. "Two Kuba Wrought-Iron Statuettes." *African Arts* 9 (1976): 60-64, 92.
Clark, J.D. *Kalambo Falls Prehistoric Site.* 2 Vols. Cambridge: Cambridge University Press, 1969 and 1974.
Cline, Walter. *Mining and Metallurgy in Negro Africa.* Menasha, Wisconsin: Banta, 1937.
Clist, Bernard. "A Critical Reappraisal of the Chronological Framework of the Early Urewe Iron Age Industry." *Muntu* 6 (1989): 35-62.
———. "Early Bantu Settlements in West-Central Africa: A Review of Recent Research." *Current Anthropology* 28, 3 (June 1987): 380-382.
——— and R. Lanfranchi, eds. *Aux Origines de l'Afrique Centrale.* Libreville: CICIBA, 1991.
Colle, P. *Les Baluba.* 2 vols. Brussels: de Wit, 1913.
Collomb, Gérard. "Quelques aspects techniques de la forge dans le bassin de l'Ogooué (Gabon)." *Anthropos* 76 (1981): 50-66.
Connah, Graham. *African Civilizations.* Cambridge: Cambridge University Press, 1987.
Conrad, David and Barbara E. Frank, eds. *Status and Identity in West Africa.* Bloomington: Indiana University Press, 1995.

Cooper, Frederick. *From Slaves to Squatters.* New Haven: Yale University Press, 1980.

———. *Decolonization and African Society: The Labor Question in French and British Africa.* Cambridge:Cambridge University Press, 1996.

Coquery-Vidrovich, Catherine. *African Women: A Modern History.* Boulder, Colorado: Westview Press, 1997.

——— and Paul Lovejoy, eds. *The Workers of African Trade.* Beverly Hills, California: Sage, 1985.

Coquilhat, Camille. *Sur le Haut-Congo.* Paris: Lebègue, 1888.

Cordell, Dennis. "The Savanna Belt of North-Central Africa." In *History of Central Africa,* 2 vols., eds. Phyllis Martin and David Birmingham. London: Longman, 1983: vol. 1, 30-75.

Cornet, Joseph. "La Société des Chasseurs d'elephants chez les Ipanga." *Annales Aequatoria* 1 (1980): 239-250.

———. *Art Royal Kuba.* Milan: Grafica Sipiel, 1982.

Cornet, Jules. "Les Gisements Métallifères du Katanga." *Bulletin de la Société Belge de Géologie* 17 (1903): 3-45.

Czekanowski, Jan. *Forschungen im Nil-Kongo-Zwischengebeit wissenschaftliche Ergebnisse der deutschen Zentral-Afrika Expedition. 1907-8.* 5 vols. Leipzig: Klinkhardt und Biermann, 1911–27.

de Barros, Philip. "Societal Repercussions of the Rise of Large-Scale Traditional Iron Production: A West African Example." *The African Archaeological Review* 6 (1988): 91-113.

de Boeck, Guy. *Baoni. Les Revoltes de la Force Publique sous Léopold II, Congo 1895-1908.* Anvers: EPO, 1987.

de Calonne-Beaufaict, A. *Les Ababua.* Brussels: Polleunis & Ceuterick, 1909.

de Heusch, Luc. "Le Symbolisme du Forgeron en Afrique." *Reflets du Monde* 10 (1956): 57-70.

de Kun, N. "La Vie et la Voyage de Ladislas Magyar dans l'intérieur du Congo en 1850-1852." *Bulletin des Séances de l'Academie Royal des Sciences d'Outre-mer* (Brussels) 6 (1960): 605-636.

de Maret, Pierre. "Ceux qui jouent avec le Feu: la Place du Forgeron en Afrique Centrale." *Africa* 50, 3 (1980): 263-279.

———. "L'Evolution Monétaire du Shaba Central entre le 7 et le 18 siècle." *African Economic History* 10 (1981): 117-149.

———. "Recent Archaeological Research and Dates from Central Africa." *Journal of African History* 26 (1985): 129-148.

———. *Fouilles Archéologiques dans la Vallée du Haut-Lualaba. II. Sanga et Katongo, 1974.* Tervuren: MRAC, 1985.

———. "Histoires de Croisettes." In *Objets-signes d'Afrique,* ed. Luc de Heusch. Tervuren: MRAC, 1995: 133-145.

——— and F. Nsuka. "History of Bantu Metallurgy: Some Linguistic Aspects." *History in Africa* 4 (1977): 43-65.

——— and G. Thiry. "How Old Is the Iron Age in Central Africa?" In *The Culture and Technology of African Iron Production,* ed. Peter Schmidt, 29-40. Gainesville: University Press of Florida, 1996.

Denbow, James. "Congo to Kalahari: Data and Hypotheses about the Political Economy of the Western Stream of the Early Iron Age." *The African Archaeological Review* 8 (1990): 139-176.

de Rode. "Note sur la fonte du fer." *Aequatoria* 3, 4 (1940): 103.

de Rop, A. "Le Forgeron Nkundo." *Annales de Notre-Dame du Sacre Coeur* (February 1956): 22-26.

de Ryck, F. *Les Lalia-Ngolu.* Anvers: Le trait d'union, 1937.

Deschamps, Hubert. *Traditions Orales et Archives au Gabon.* Paris: Berger-Levrault, 1962.

de Sousberghe, Léon. "Forgerons et Fondeurs de Fer chez les Ba-Pende et leurs Voisins." *Zaire* 9, 1 (1955): 25-31.

de Waal, Alex. "The Genocidal State." *The Times Literary Supplement* (July 1, 1994): 3-4.

Digombe, L. P. Schmidt, V. Mouleingui-Boukosso, J.-B. Mombo, and M. Locko. "Gabon: The Earliest Iron Age of West Central Africa." *Nyame Akuma* 28 (April 1987): 9-11.

Douglas, Mary. "Lele Economy compared with the Bushong." In *Markets in Africa,* eds. Paul Bohannan and George Dalton, 211-237. Evanston: Northwestern University Press, 1962.

Dupré, Georges. *Un Ordre et sa Destruction. Anthropologie des Nzabi.* Paris: ORSTOM, 1982.

———. "The History and Adventures of a Monetary Object of the Kwélé of the Congo: Mezong, Mondjos, and Mondjong." In *Money Matters: Instability, Values and Social Payments in the Modern History of West African Communities*, ed. Jane Guyer, pp. 77-97. Portsmouth, NH: Heinemann, 1995.

Dupré, Marie-Claude. "Pour une Histoire des Productions: la Métallurgie du fer chez les Téké Ngungulu, Tio, Tsaayi (République du Congo)." *Cahiers ORSTOM* 18 (1981-2): 195-223.

———. "L'outil Agricole des Essartages Forestiers; le Couteau de Culture au Gabon et au Congo." Unpublished ms., 1990.

——— and Bruno Pinçon. *Métallurgie et Politique en Afrique Centrale.* Paris: Karthala, 1997.

Dye, Eva. *Bolenge. The Story of Gospel Triumphs on the Congo.* Cincinnati, Ohio: Foreign Christian Missionary Society, 1910.

Eggert, Manfred. "Katuruka und Kemondo: zur Komplexität der frühen Eisentechnik in Afrika." *Beiträge zur allgemeinen und vergleichenden Archäologie* 7 (1985): 243-263.

———. "Aspects de l'Ethnohistoire Mongo: Une vue d'ensemble sur les Populations de la Rivière Ruki." *Annales Aequatoria* 1 (1980): 149-168.

———. "On the alleged Complexity of Early and Recent Iron Smelting in Africa: Further Comments on the Preheating Hypothesis." *Journal of Field Archaeology* 14 (1987): 377-382.

———. "Historical Linguistics and Prehistoric Archaeology: Trend and Pattern in Early Iron Age Research of sub-Saharan Africa." *Beiträge zur allgemeinen und vergleichenden Archäologie* 3 (1981): 277-324.

———. "The Current State of Archaeological Research in the Equatorial Rainforest of Zaire." *Nyame Akuma* 24/25 (December 1984): 39-42.

———. "Remarks on Exploring Archaeologically Unknown Rain Forest Territory: The Case of Central Africa." *Beiträge zur allgemeinen und vergleichenden Archäologie* 5 (1983): 283-322.

———. "Imbongo and Batalimo: Ceramic Evidence for Early Settlement of the Equatorial Rainforest." *African Archaeological Review* 5 (1987): 129-145.

Eggert, R.K. "Zur Rolle des Wertmessers (*mitako*) am oberen Zaire, 1877-1908." *Annales Aequatoria* 1 (1980): 263-324.

Einzig, Paul. *Primitive Money.* 2nd ed. Oxford: Pergamon Press, 1966.

Eliade, Mircea. *The Forge and the Crucible.* New York: Harper & Row, 1971. (English translation by Rider and Company [1962] of 1956 French edition.)

Engels, A. *Les Wangata. Etude Ethnographique.* Brussels: Vromant, 1912.

———. "Les Wangata." *La Revue Congolaise* 1 (1910).

Estermann, Carlos. *The Ethnography of Southwestern Angola.* 3 vols. New York: Africana Publishing Co., 1976.

Fagan, Brian. "Excavations at Ingombe Ilede 1960-2." In *Iron Age Cultures in Zambia.* 2 vols., ed. B. Fagan, D.W. Phillipson, and S.G.H. Daniels. London: Chatto & Windus, 1967 and 1969; vol. 2: 55-184.

Fairley, Nancy. "Mianda ya Ben'Ekie: A History of the Ben'Ekie." Ph.D. Dissertation, State University of New York at Stony Brook, 1978.

Foquet-Vanderkerken. "Les Populations Indigènes des Territoires de Kutu et de Nseontin." *Congo. Revue Générale de la Colonie Belge* 5, 2 (1924): 129-171.

Freund, Bill. *The African Worker.* Cambridge: Cambridge University Press, 1988.

Frobenius, Leo. "Afrikanische Messer." *Prometheus* 620 (1901): 753-759.

Galbraith, John. *Money. Whence it Came, Where it Went.* Boston: Houghton Mifflin, 1975.

Gann, L.H. and Peter Duignan. *The Rulers of Belgian Africa 1884-1914.* Princeton: Princeton University Press, 1979.

Glave, E.J. *Six years of Adventure in Congo-land.* London: Sampson Low, Marston & Co., 1893.

Gordon, Robert and David Killick. "Adaptation of Technology to Culture and Environment: Bloomery Iron Smelting in America and Africa." *Technology and Culture* 34, 2 (1993): 243-270.

Goucher, Candice. "Iron is Iron 'til it is Rust: Trade and Ecology in the Decline of West African Iron Smelting." *Journal of African History* 22 (1981): 179-189.

Guillot, B. "Note sur les anciennes mines de fer du pays Nzabi dans la région de Mayoko." *Cahiers ORSTOM* 6, 2 (1969): 93-99.

Guiral, Léon. *Le Congo Français.* Paris: E. Plon, 1889.

Guthrie, Malcolm. *Comparative Bantu.* 4 vols. Farnborough, England: Gregg, 1969-70.

Guyer, Jane. "Indigenous Currencies and the History of Marriage Payments." *Cahiers d'Etudes Africaines* 104, 26-4 (1986): 577-610.

———, ed. *Money Matters. Instability, Values and Social Payments in the Modern History of West African Communities.* Portsmouth, NH: Heinemann, 1995.

Hagendorens, J. *Dictionnaire Otetela-Français.* Bandundu: CEEBA, 1975.

Halkin, Joseph. *Les Ababua.* Brussels: de Wit, 1911.

Harms, Robert. *River of Wealth, River of Sorrow: The Central Zaire Basin in the Era of the Slave and Ivory Trade 1500-1891.* New Haven: Yale University Press, 1981.

Harries, Patrick. *Work, Culture, and Identity.* Portsmouth, NH: Heinemann, 1994.

Hawker, George. *The Life of George Grenfell.* New York: Fleming Revel Co., 1909.

Herbert, Eugenia. *Iron, Gender, and Power. Rituals of Transformation in African Societies.* Bloomington: Indiana University Press, 1993.

Hersak, Dunja. *Songye Masks and Figure Sculpture.* London: Ethnographica, 1986.

Heywood, Linda. "Production, Trade and Power—the Political Economy of Central Angola 1850–1930." Ph.D. Dissertation, Columbia University, 1984.

Hiernaux, Jean, Emma de Longrée, and Josse de Buyst. *Fouilles Archéologiques dans la Vallée du Haut-Lualaba. I. Sanga, 1958.* Tervuren: MRAC, 1971.

Hilton, Anne. *The Kingdom of Kongo.* Oxford: Clarendon, 1985.
Hinde, Sidney L. *The Fall of the Congo Arabs.* 1897. Reprint, New York: Negro Universities Press, 1969.
Hobsbawm, Eric J. *Labouring Men. Studies in the History of Labour.* London: Weidenfield & Nicolson, 1964.
———. *Workers: Worlds of Labor.* New York: Pantheon Books, 1984.
Hulstaert, Gustave. *Les Mongo. Aperçu Général.* Tervuren: Musée Royale de l'Afrique Centrale, 1961.
———. "Aux origines de Mbandaka." In *Histoire Ancienne de Mbandaka,* eds. D. Vangroenweghe, G. Hulstaert, and L. Lufungula, 75-147. Mbandaka: Centre Aequatoria, 1986.
———. *Proverbes Mongo.* Tervuren: Musée du Congo Belge, 1958.
———. *Notes de Botanique Mongo.* Brussels: ARSOM, 1966.
———. *Dictionnaire Lomongo-Français.* Tervuren: Musée Royale du Congo Belge, 1957.
———. "Notes sur le Mariage des Ekonda." *Aequatoria* 11 (Dec. 1938): 1-11.
———. *Le Mariage des Nkundo.* Brussels: 1938.
Hultgren, Mary Lou and Jeanne Zeidler. *A Taste for the Beautiful.* Hampton, Virginia: Hampton University Museum, 1993.
Isaacman, Allen. *Cotton is the Mother of Poverty.* Portsmouth, NH: Heinemann, 1996.
——— and Jan Vansina. "African Initiatives and Resistance in Central Africa, 1880-1914." In *UNESCO General History of Africa,* ed. Adu Boahen, 8 vols. Paris: UNESCO, 1985: vol. 7, 169-194.
Jameson, James. *The Story of the Rear Column of the Emin Pasha Relief Expedition.* Toronto: Rose, 1891.
Jeannest, Charles. *Quatre années au Congo.* Paris: Charpentier, 1883.
Jewsiewicki, Bogumil. "Belgian Africa." In *The Cambridge History of Africa,* General eds. J. D. Fage and Roland Oliver, 8 vols. Cambridge: Cambridge University Press, 1986: vol. 7, 460-494.
———. "Rural Society and the Belgian Colonial Economy." In *History of Central Africa,* eds. Phyllis Martin and David Birmingham. 2 vols. London: Longman, 1983: vol. 2, 95-126.
Johnston, H.H. *George Grenfell and the Congo.* 2 vols. London: Hutchinson, 1908.
———. *The River Congo from its Mouth to Bolobo.* 1884. Reprint, Detroit: Negro University Press, 1970.
Junker, Wilhelm. *Travels in Africa during the Years 1875-1886.* 3 vols. London: Chapman & Hall, 1890–92.
Keesman, Ingo, Johannes Preuss, and Johannes Endres. "Eisengewinnung aus laterischen Erzen, Ruki-Region, Provinz Equator/Zaire." *Offa* 40 (1983): 183-190.
Keim, Curtis. "Precolonial Mangbetu Rule: Political and Economic Factors in Nineteenth Century Mangbetu History (Northeast Zaire)." Ph.D. Dissertation, Indiana University, 1979.
Killick, David J. "Technology in its Social Setting: Bloomery Iron Smelting at Kasungu, Malawi, 1860–1940." Ph.D. Dissertation, Yale University, 1990.
———. "On Claims for 'Advanced' Ironworking Technology in Precolonial Africa." In *The Culture and Technology of African Iron Production,* ed. Peter Schmidt, 247-267. Gainesville: University Press of Florida, 1996.
———. "The Relevance of Recent Iron-Smelting Practice to Reconstructions of Prehistoric Smelting Technology." *MASCA Research Papers in Science and Technology* 8, 1 (1991): 47-54.

Klein, Hildegard, ed. *Leo Frobenius: Ethnographische Notizen aus den Jahren 1905 und 1906.* Wiesbaden: Steiner, 1985–90.

Koloss, Hans-Joachim. *Art of Central Africa.* New York: Metropolitan Museum of Art, 1990.

Korse, Piet. "La Forge." *Annales Aequatoria* 9 (1988): 23-35.

Krieger, Kurt. *Westafrikanische Plastik.* 3 Vols. Berlin: Museum für Völkerkunde, 1965.

Kriger, Colleen. "Ironworking in 19th Century Central Africa." Ph.D. Dissertation, York University (Canada), 1992.

———. "Museum Collections as Sources for African History." *History in Africa* 23 (1996): 129-154.

———. "A Critical Look at Mircea Eliade and the Myth of the Mystical Blacksmith." Paper presented at the African Studies Association annual meeting, Columbus, Ohio, November 1997.

———. "The Knife and the Hoe: Gender, Metals and Sources of Power in the Middle Zaire Basin, late nineteenth and early twentieth century." Paper presented at conference on Technology and Gender, Oxford, May 1993.

Kubler, George. *The Shape of Time.* New Haven: Yale University Press, 1962.

Kyankinge Masandi Kita. "La Technique Traditionelle de la Métallurgie du fer chez les Balega de Pangi (Zaire)." *Muntu* 3 (1985): 85-98.

Laman, Karl. *The Kongo.* 4 Vols. Uppsala: Studia Ethnographica, 1953–68.

LaViolette, Adria. "An Archaeological Ethnography of Blacksmiths, Potters, and Masons in Jenne, Mali (West Africa)." Ph.D. Dissertation, Washington University, 1987.

Lemaire, Charles. *Congo et Belgique.* Brussels: Bulens, 1894.

———. "Une forge à l'Equateur." *Le Congo Illustré* (1892): 168.

———. *Au Congo. Comment les Noirs Travaillent.* Brussels: Bulens, 1895.

Leyder, Jean. "Notes sur le couteau 'ngwolo' des Ngombe de la Lulonga (Congo Belge)." *2e Congrès National des Sciences, comptes rendus (Brussels)* II (1935): 937-946.

Liegeois, L. "Notice sur le régime social des Basongo Meno de Kole." *Bulletin des Juridictions Indigènes et du Droit Coutumier Congolais* 9, 1 (1941): 13-23.

Lima, Augusto Guilhermo Mesquitela. "Le Fer in Angola." *Cahiers d'Etudes Africaines* 17, 2-3 (1966–67): 345-351.

Lumbwe Mudindaambi. *Dictionnaire Mbala-Français.* 4 vols. Bandundu: CEEBA, 1977-81.

———. *Objets et Techniques de la vie Quotidienne Mbala.* 2 vols. Bandundu: CEEBA, 1976.

McIntosh, Susan Keech and Roderick J. McIntosh. "From Stone to Metal: New Perspectives on the Later Prehistory of West Africa." *Journal of World Prehistory* 2, 1 (1988): 89-133.

McMaster, Mary Allen. "Patterns of Interaction: A Comparative Ethnolinguistic Perspective on the Uele Region of Zaire ca. 500 B.C. to 1900 A.D." Ph.D. dissertation, University of California at Los Angeles, 1988.

McNaughton, Patrick R. *Secret Sculptures of Komo.* Philadelphia: Institute for the Study of Human Issues, Inc., 1979.

———. *The Mande Blacksmiths.* Bloomington: Indiana University Press, 1988.

Maes, Joseph. "La Métallurgie chez les Populations de Lac Léopold II-Lukenie." *Ethnologica* 4 (1930): 68-101.

Magyar, László. *Reisen in süd-Afrika in den Jahren 1849 bis 1857.* 1859. Reprint, Nendeln: Kraus Reprint.

Mahieu, Adolphe. "Numismatique du Congo." *Congo. Revue Générale de la Colonie Belge* 1 (1923): 1-55.

Mandala, Elias. *Work and Control in a Peasant Economy.* Madison: University of Wisconsin Press, 1990.

Maquet, Emma. *Outils de Forge du Congo, du Rwanda, et du Burundi.* Tervuren: Musée Royale de l'Afrique Centrale, 1965.

Marcadé, Bernard. "Chérie Samba." *Galeries Magazine* (February/March 1991): 84-87.

Marchal, Jules. *E.D. Morel contre Léopold II.* 2 vols. Paris: l'Harmattan, 1996.

Masui, T. *Guide de la Section de l'Etat Indépendant du Congo à l'Exposition de Bruxelles-Tervuren en 1897.* Bruxelles: Veuve Monnom, 1897.

Mertens, Joseph. "Les BaDzing de la Kamtsha." *Mémoires. Institut Royal colonial Belge, Section des Sciences Morales et Politiques* 4 (1935).

Miller, Duncan and Nikolaas Van Der Merwe. "Early Metalworking in sub-Saharan Africa: A Review of Recent Research." *Journal of African History* 35 (1994): 1-36.

Miller, Joseph. *Kings and Kinsmen. Early Mbundu States in Angola.* Oxford: Clarendon, 1976.

———. *Slavery, a Worldwide Bibliography, 1900-1982.* White Plains, NY: Kraus International, 1985.

———. *Slavery and Slaving in World History: A Bibliography, 1900-1991.* Millwood, NY: Kraus International, 1993.

———. "Cokwe Trade and Conquest in the Nineteenth Century." In *Precolonial African Trade,* eds. Richard Gray and David Birmingham. London: Oxford University Press, 1970: 175-202.

———. "The Paradoxes of Impoverishment in the Atlantic Zone." In *History of Central Africa,* eds. Phyllis Martin and David Birmingham. 2 vols. London: Longman, 1983: vol. 1, 118-160.

Miracle, Marvin. *Agriculture in the Congo Basin.* Madison: University of Wisconsin Press, 1967.

"La Monnaie." *Le Congo Illustré* 9 (1892): 34-35.

Monteiro, Joachim. *Angola and the River Congo.* 2 vols. New York: Macmillan, 1876.

Moorhead, Max, ed. *Missionary Pioneering in Congo Forests.* Preston, England: Seed & Sons, 1922.

Mumbanza mwa Bawele na Nyabakombi Ensobato. "Les Forgerons de la Ngiri. Une Elite Artisanale parmi les Pêcheurs." *Enquêtes et Documents d'Histoire Africaine* 4 (1989): 114-132.

Munro, J. Forbes. *Africa and the International Economy 1800-1960.* London: Dent & Sons, Ltd., 1976.

Nahan, P. "Reconnaissance de Banalia vers Buta et Retour à Bolulu." *Belgique Coloniale* 4 (1898): 544-546, 557-559.

Nelson, Samuel H. *Colonialism in the Congo Basin, 1880-1940.* Athens, Ohio: Ohio University Press, 1994.

Nenquin, Jacques. *Excavations at Sanga, 1957.* Tervuren: Musée Royale de l'Afrique Centrale, 1963.

Oliver, Roland and Anthony Atmore. *The African Middle Ages 1400-1800.* Cambridge: Cambridge University Press, 1989.

Osumaka Likaka. *Rural Society and Cotton in Colonial Zaire.* Madison: University of Wisconsin Press, 1997.

Parpart, Jane. "The Labor Aristocracy Debate in Africa." *African Economic History* 13 (1984): 171-191.

Phillipson, D.W. *The Later Prehistory of Eastern and Southern Africa.* London: Heinemann, 1977.

———. "Cewa, Leya, and Lala Iron-Smelting Furnaces." *South African Archaeological Bulletin* 23 (1968): 102-113.

——— and Brian Fagan. "The Date of the Ingombe Ilede Burials." *Journal of African History* 10, 2 (1969): 199-204.

Pole, L.M. "Decline or Survival? Iron Production in West Africa from the Seventeenth to the Twentieth Centuries." *Journal of African History* 23 (1982): 503-513.

Pruitt, William, Jr. "An Independent People: A History of the Sala Mpasu of Zaire and their Neighbors." Ph.D. Dissertation, Northwestern University, 1973.

Quiggin, A. Hingston. *A Survey of Primitive Money.* 1949. Reprint, New York: Barnes & Noble, 1970.

Quirin, James. *The Evolution of the Ethiopian Jews.* Philadelphia: University of Philadelphia Press, 1992.

Read, Frank. "Iron Smelting and Native Blacksmithing in Ondulu Country, Southeast Angola." *African Affairs* 2, 5 (1902–3): 44-49.

Redinha, Jose. *Campanha Etnografica ao Tchiboco.* Lisboa: CITA, 1953.

Reefe, Thomas. *The Rainbow and the Kings: History of the Luba Empire to 1891.* Los Angeles: University of California Press, 1981.

———. "The Societies of the Eastern Savanna." In *History of Central Africa,* eds. Phyllis Martin and David Birmingham. 2 vols. London: Longman, 1983: vol. 1, 160-205.

Rehder, J.E. "Use of Preheated Air in Primitive Furnaces: Comment on Views of Avery and Schmidt." *Journal of Field Archaeology* 13 (1986): 351-353.

———. "Iron versus Bronze for Edge Tools and Weapons: A Metallurgical View." *Journal of Metals* (August, 1992): 42-46.

———. "Primitive Furnaces and the Development of Metallurgy." *Journal of the Historical Metallurgy Society* 20, 2 (1986): 87-92.

Roberts, Andrew. *A History of the Bemba.* Madison: University of Wisconsin Press, 1973.

Robinson, Ian G. "Hoes and Metal Templates in Northern Cameroon." In *An African Commitment,* eds. Judy Sterner and Nicholas David. Calgary: University of Calgary Press, 1992: 231-241.

Robinson, K.R. *Khami Ruins.* Cambridge: Cambridge University Press, 1959.

Rombauts, H. "Ekonda e Mputela." *Aequatoria* 9, 4 (1946): 138-152.

———. "Les Ekonda." *Aequatoria* 8, 4 (1945): 121-127.

Rosen, Miriam. "Chérie Samba, Griot of Kinshasa and Paris." *Artforum* (March, 1990): 137-140.

Rybczynski, Witold. "The Mystery of Cities." *New York Review of Books* (15 July 1993): 13-16.

Schildkrout, Enid and Curtis Keim. *African Reflections. Art from Northeast Zaire.* Seattle: University of Washington Press, 1990.

Schmidt, Peter, ed. *The Culture and Technology of African Iron Production.* Gainesville: University Press of Florida, 1996.

——— and Donald Avery. "Complex Iron Smelting and Prehistoric Culture in Tanzania." *Science* 201, 4361 (22 September 1978): 1085-1089.

Schoonheyt, J.A. "Les Croisettes du Katanga." *Revue Belge de Numismatique* 137 (1991): 141-157.

Schultz, Ronald. "The Small-Producer Tradition and the Moral Origins of Artisan Radicalism in Philadelphia, 1720-1810." *Past and Present* 127 (May, 1990): 84-116.

Schwartz, D., R. Deschamps, and M. Fournier. "Un site de fonte du fer Récent (300 bp) et original dans le Mayombe congolais: Ganda-Kimpese." *Nsi* 8/9 (1991): 33-40.

Schweinfurth, Georg. *Artes Africanae. Illustrations and Descriptions of Productions of the Industrial Arts of Central African Tribes.* Leipzig: Brockhaus, 1875.

———. *The Heart of Africa: Three Years' Travels and Adventures in the Unexplored Regions of Central Africa from 1868 to 1871.* 2 vols. New York: Harper & Bros., 1874.

Scott, Joan Wallach. *The Glassworkers of Carmaux.* Cambridge: Harvard University Press, 1974.

Sheppard, William. "Into the Heart of Africa." *Southern Workman* (April, 1895): 61-66.

Smail, John. "Manufacturer or Artisan? The Relationship between Economic and Cultural Change in the Early Stages of the Eighteenth-century Industrialization." *Journal of Social History* 25, 4 (1992): 791-814.

Smith, Cyril Stanley. "Metallurgy in the Seventeenth and Eighteenth Centuries." In *Technology in Western Civilization,* eds. Melvin Kransberg and Carroll Pursell, vol. 1, 142-167. New York: Oxford University Press, 1967.

Stanley, Henry Morton. *Through the Dark Continent.* 2 vols. Toronto: Magurn, 1878.

———. *In Darkest Africa.* 2 vols. New York: Scribners, 1891.

Starr, Frederick. *Congo Natives.* Chicago: Lakeside Press, 1912.

Stas, J.B. "Les Nkumu chez les Ntomba de Bikoro." *Aequatoria* 2, 10-11 (Oct.-Nov. 1939): 109-123.

Sura Dji. Visages et Racines du Zaire. Paris: Musée des Arts Décoratifs, 1982.

Sutton, J.E.G. "Temporal and Spatial Variability in African Iron Furnaces." In *African Ironworking—Ancient and Traditional,* eds. R. Haaland and P. Shinnie, 164-196. Oslo: Norwegian Universities Press, 1985.

Tamari, Tal. "The Development of Caste Systems in West Africa." *Journal of African History* 32 (1991): 221-250.

Thompson, E.P. *The Making of the English Working Class.* New York: Vintage, 1963.

Thonner, Franz. *Dans la Grande Forêt de l'Afrique Centrale.* Brussels: Société Belge de Librairie, 1899.

Thornton, John. *The Kingdom of Kongo: Civil War and Transition, 1641–1718.* Madison: University of Wisconsin Press, 1983.

———. *Africa and Africans in the Making of the Atlantic World, 1400–1680.* Cambridge: Cambridge University Press, 1992.

Timmermans, P. "Les Sapo Sapo près de Luluabourg." *Africa-Tervuren* 8, 1-2 (1962): 29-53.

———. "Voyage à travers le Gabon et le sud-ouest du Congo-Brazzaville." *Africa-Tervuren* 10, 3 (1964): 69-78.

Torday, Emil and T.A. Joyce. *Notes Ethnographiques sur les Populations habitant les Bassins du Kasai et du Kwango Oriental.* Tervuren: Musée du Congo Belge, 1922.

———. *Notes Ethnographiques sur les Peuples communément appelés Bakuba, ainsi que sur les Peuplades Apparentées, les Bushongo.* Tervuren: Musée du Congo Belge, 1910.

Trigger, Bruce. *A History of Archaeological Thought.* Cambridge: Cambridge University Press, 1989.

Tylecote, R.F. *A History of Metallurgy.* London: The Metals Society, 1976.

———. "Furnaces, Crucibles, and Slags." In *The Coming of the Age of Iron,* eds. T. Wertime and J. Muhly, 183-229. New Haven: Yale University Press, 1980.

———, J.N. Austin, and A.E. Wraith. "The Mechanism of the Bloomery Process in Shaft Furnaces." *Journal of the Iron and Steel Institute* 209 (May 1971): 242-363.

Untracht, Oppi. *Metal Techniques for Craftsmen*. London: Robert Hale, Ltd., 1969.

Van der Kerken, G. *l'Ethnie Mongo.* 2 vols. Brussels: ARSOM, 1944.

Vangroenweghe, Daniel. *Bobongo: La grande fête des Ekonda.* Berlin: Dietrich Reimer, 1988.

Van Grunderbeek, M.-C., E. Roche, and H. Doutrelepont. "L'Age du Fer Ancien au Rwanda et au Burundi. Archéologie et Environnement." *Journal des Africanistes* 52, 1-2 (1982): 5-58.

Van Noten, Francis. "Ancient and Modern Iron Smelting in Central Africa: Zaire, Rwanda, and Burundi." In *African Iron Working—Ancient and Traditional,* eds. R. Haaland and P. Shinnie, 102-120. Oslo: Norwegian Universities Press, 1985.

———. *The Archaeology of Central Africa*. Graz: Akademische Druck– und Verlagsanstalt, 1982.

——— and E. Van Noten. "Het Ijzersmelten bij de Madi." *Africa-Tervuren* 20, 3-4 (1974): 57-66.

van Overbergh, C. and E. de Jonghe. *Les Mangbetu.* Brussels: de Wit, 1909.

Vansina, Jan. *The Tio Kingdom of the Middle Congo, 1880-1892.* London: Oxford University Press, 1973.

———. *Introduction à l'Ethnographie du Congo.* Kinshasa: Université Lovanium, 1966.

———. *Le Royaume Kuba.* Tervuren: Musée Royale de l'Afrique Centrale, 1964.

———. *Geschiedenis van de Kuba.* Tervuren: Musée Royale de l'Afrique Centrale, 1963.

———. "Trade and Markets among the Kuba." In *Markets in Africa,* eds. Paul Bohannan and George Dalton, 190-211. Evanston: Northwestern University Press, 1962.

———. "Peoples of the Forest" In *History of Central Africa*, eds. Phyllis Martin and David Birmingham. 2 vols. London: Longman, 1983: vol. 1, 75-118.

———. "The Bells of Kings." *Journal of African History* 10, 2 (1969): 187-197.

———. *Kingdoms of the Savanna.* Madison: University of Wisconsin Press, 1968.

———. *Children of Woot.* Madison: University of Wisconsin Press, 1978.

———. *Paths in the Rainforests.* Madison: University of Wisconsin Press, 1990.

———. "Bantu in the Crystal Ball, parts I and II." In *History in Africa* 6 (1979): 287-293, and 7 (1980): 293-325.

———. "Lignage, Ideologie, et Histoire en Afrique Equatoriale." *Enquêtes et Documents d'Histoire Africaine* 4 (1980): 133-155.

———. "New Linguistic Evidence and 'the Bantu Expansion'." *Journal of African History* 36 (1995): 173-195.

Védy, L. "Les Ababuas." *Bulletin de la Société Royale belge de Géographie* 28 (1904): 191-205.

Vellut, Jean-Luc. "Ethnicity and Genocide in Rwanda." *The Times Literary Supplement* (July 15, 1994): 17.

———. "Notes sur le Lunda et la Frontière Luso-Africaine (1700-1900)." *Etudes d'Histoire Africaine* 3 (1972): 61-166.

———. "La Violence Armée dans l'Etat Indépendant du Congo." *Cultures et Développement* 16, 3-4 (1984): 671-707.

———. "The Congo Basin and Angola." In *UNESCO General History of Africa,* ed. J. F. Ade Ajayi. 8 vols. Paris: UNESCO, 1989: vol. 7, 294-325.

———. "Mining in the Belgian Congo." In *History of Central Africa,* eds. Phyllis Martin and David Birmingham. 2 vols. London: Longman, 1983: vol. 2, 126-163.

Verner, S. *Pioneering in Central Africa.* Richmond, Virginia: Presbyterian Committee of Publication, 1903.

Volavka, Zdenka. *Crown and Ritual: Royal Insignia of Ngoyo.* Toronto: University of Toronto Press, 1998.

von François, Kurt. *die Erforschung des Tschuapa und Lulongo.* Leipzig: Brockhaus, 1888.

von Wissmann, Hermann, Ludwig Wolf, Kurt von François, and Hans Mueller. *Im innern Afrikas.* 1888. Reprint, Nendeln: Kraus Reprint, 1974.

———. and Paul Pogge. *Unter deutsche Flagge quer durch Afrika.* Berlin: Walther & Apoplant, 1890.

Ward, Herbert. *Among Congo Cannibals.* 1891. Reprint, New York: Negro Universities Press, 1969.

Warnier, Jean-Pierre and Ian Fowler. "A Nineteenth Century Ruhr in Central Africa." *Africa* 49, 4 (1979): 329-351.

Washburn, Dorothy. "Style, Classification, and Ethnicity: Design Categories on Bakuba Raffia Cloth." *Transactions of the American Philosophical Society* 80, Pt. 3 (1990).

Wauters, G. "L'Institution du 'nkum'." *Annales des Missionaires du Sacre Coeur* (1937): 152-160.

Webb, James L.A., Jr. "Toward the Comparative Study of Money: A Reconsideration of West African Currencies and Neoclassical Monetary Concepts." *International Journal of African Historical Studies* 15, 3 (1982): 455-466.

White, Luise. *The Comforts of Home: Prostitution in Colonial Nairobi.* Chicago: University of Chicago Press, 1990.

Whitehead, John. *Grammar and Dictionary of the Bobangi Language.* London: Baptist Missionary Society and Kegan Paul, Trench Trübner & Co. Ltd. 1899.

Womersley, Harold. *Legends and History of the Luba.* Los Angeles: Crossroads Press, 1984.

Woodhouse, H.C. "Elephant Hunting by Hamstringing Depicted in the Rock Paintings of Southern Africa." *South African Journal of Science* 72, 6 (June 1976): 175-177.

Wright, Marcia. "Iron and Regional History: Report on a Research Project in Southwestern Tanzania." *African Economic History* 14 (1985): 147-165.

Yoder, John C. *The Kanyok of Zaire.* Cambridge: Cambridge University Press, 1992.

Index

About the Author

COLLEEN E. KRIGER is Assistant Professor of History at the University of North Carolina, Greensboro.